Texas A&M University Military History Series

The American Home Guard

THE STATE MILITIA IN THE TWENTIETH CENTURY

Barry M. Stentiford

Texas A&M University Press • College Station

First edition

The paper used in this book meets the minimum requirements
of the American National Standard for Permanence
of Paper for Printed Library Materials, Z39.48-1984.
Binding materials have been chosen for durability.

Library of Congress Cataloging-in-Publication Data

Stentiford, Barry M.
The American home guard : the state militia in the twentieth century / Barry M. Stentiford.—1st ed.
p. cm.—(Texas A&M University military history series ; 78)
Includes bibliographical references and index.
ISBN 1-58544-181-3 (cloth : alk. paper)
1. United States—Militia—History—20th century.
I. Title. II. Texas A&M University military history series ; 78.
UA42.S78 2002
355.3'7'0973—dc21 2001006543

For my wife,

VITIDA

Contents

Acknowledgments

I owe a great deal of gratitude to Professor Harold Selesky of the University of Alabama, who directed my doctoral dissertation, which formed the basis for this book, and who shared my enthusiasm for state militia. Dr. Selesky told me what questions to ask and carefully read some very rough drafts to tell me what worked and what did not. The Graduate Council at the University of Alabama granted me the research fellowship that allowed me to research the subject properly. Maj. (ret.) Thomas M. Weaver, the former librarian at the National Guard Association of the United States headquarters in Washington, D.C., provided valuable support and assistance for me during long weeks in his stacks.

I also owe thanks to my family, especially my brother, Tim, and my sister, Joan, who by their example took away all my excuses for not going to college. I need to thank my sister and her husband, Gary Ulmer, who patiently allowed me to freeload for several weeks at their house while researching.

And last, I must thank my wife, Vitida, who inspires me every day.

Note on Terms

The terminology in this study can lead to confusion because many of the terms have common usages that differ from their precise meaning. In addition, many terms have a slightly different meaning depending on context and year. What follows is an explanation of most of the ambiguous terms in this study.

Army Reserve The modern Army Reserve has three parts: the Ready Reserve, the Retired Reserve, and the Standby Reserve. Drilling units of the Army Reserve (Troop Program Units or "TPUs"), are part of the Ready Reserve. They were formerly known as the "Organized Reserve," and are similar to the National Guard, but have no state status. Members are not normally eligible for militia service, although in the period before the United States entered World War II, the War Department allowed reserve officers not on active duty to accept a commission in the State Guard.

cadre force A unit with only the leadership positions filled. In theory, the ranks would be recruited or conscripted during mobilization.

federalization When referring to the National Guard, it means when a unit has been inspected and accepted by federal inspectors and is then part of the National Guard. Informally, the term is sometimes used to refer to bringing a National Guard unit onto active federal service.

home guard An organized militia of a town, city, county, or state, without a federal obligation. These units are usually only liable for service within the jurisdiction of the government that created them.

Home Guard Used in the official name of the organized militia forces of several states during the Great War. Usually, these forces had a statewide obligation and depended at least in part on federal assistance for material and weapons.

militia A body of civilian adult free males with an obligation to serve as a military force within the territory of the state or nation for limited periods when ordered by the state or federal government. Under modern U.S. law, women in the National Guard also belong to their state's militia.

National Guard That part of the state military forces with a dual obligation to serve both as state militia and as reserve of the Army or Air Force.

naval militia That part of the organized militia geared toward operations on the water. Usually, members of a state's naval militia are members of the U.S. Navy Reserve who accept an additional state status. Few states maintain a naval militia.

organized militia Part-time civilian military units having sanction under state laws. This includes National Guard, naval militia, and State Guard units.

Reserve Components Also called "Civilian Components." The National Guard and the reserves of the Army, Navy, Air Force, Marines, and Coast Guard taken together.

State Defense Forces A name for organized militia forces not part of the National Guard increasingly used since the 1970s. Another name for the State Guard.

State Guard The official name of the organized militia of several states during the Great War. Usually these forces had a statewide obligation and depended on the local and state resources rather than federal. Since World War II, the term has meant organized militia forces of the states without a federal mission. It has been used in the official names of the wartime militia of most states and as a generic term for such exclusively state forces.

state militia Any of the part-time civilian military forces of a state. This includes the National Guard and any other organized militia a state may maintain or organize.

unorganized militia The mass of civilian adult free males that does not belong to the National Guard, the federal military, or any organized militia force.

THE AMERICAN HOME GUARD

Introduction

The bombing of the Alfred P. Murrah Federal Building in Oklahoma City in the spring of 1995 brought to the general public a new image of "militia." Media coverage following the attack exposed the existence of hundreds of private, part-time armies throughout the nation calling themselves militia, some with racist overtones, but almost all with a strong anti-federal government ideology. Until then, the term "militia" had become almost quaint, bringing with it memories of celebrations from the American Revolution Bicentennial featuring modern-day militiamen parading with flintlock muskets, wearing tricornered hats and knickers. Serious inquiry into the modern descendants of the ancient concept of militia usually led to the National Guard or the Reserves, both of which claimed for themselves the title as heir to America's militia tradition. In the aftermath of the bombing, with news reports of private groups of citizens armed with military-style weapons training for combat against government forces, legitimate militia groups, chartered by state governments, faced a public-relations nightmare.

Public confusion and fear over militia has been fueled by a lack of understanding in the media as well as in the general population over what exactly constitutes a militia. The new image, given form by guests on daytime talk shows or on the cover of magazines, was one of groups of predominantly white men, wearing a combination of military surplus and hunting gear, seemingly obsessed with their right to bear arms under the Second Amendment to the U.S. Constitution.[1] But other images of a militia exist side by side with this new one. National Guard recruiting campaigns from the 1970s invited would-be citizen-soldiers to "be one of the new Minutemen," while Credence Clearwater Revival sang that at a concert, "someone got excited, had to call the State Militia."[2] Added to this cultural baggage was the propensity of southern men of standing to adopt the title of "colonel," even if their rank rested on limited or no martial exploits. Somewhere in the public consciousness was evidence that the militia was not a dead institution.

The most common perception was that the militia of the Revolutionary era had evolved into the modern National Guard. This perception is largely true. However, after the rise of the National Guard in the late nineteenth and early twentieth centuries, states have created, maintained, deployed, and disbanded countless militia organizations that were never part of the National Guard. These militia units, variously called home guards, State Guard, naval militia, National Guard Reserve, and State Defense Forces, have been largely overlooked by historians and forgotten by the public.[3] This oversight leaves a serious gap in the historical record of the modern militia, for it ignores the roles state militias have played during wartime, as well as the difficulties caused by the dual federal and state standings of the National Guard.

Published works on non-National Guard state militia in the twentieth century are confined mostly to a handful of articles on particular units, while books written by members have appeared on the militia in Texas, Puerto Rico, Indiana, and Michigan.[4] In addition, a few state military departments have written histories of their wartime militia.[5] The surviving organizational badges of these militia organizations have inspired a few collectors to specialize in such relics of forgotten units.[6] In at least one state, a former member of a wartime militia self-published a history of his force during the Great War in hopes of gaining some recognition for the men who served with him.[7]

What follows is the story of these militia forces.[8] They have remained outside of the National Guard, yet at the same time their history has been intimately intertwined with that of the National Guard. Ironically, the very existence of the National Guard has made the creation of other militia forces necessary.[9] Modern state militiamen, heir to an ancient tradition, struggle to define a role for themselves in a society which increasingly views them as anachronistic.

CHAPTER 1

Origins of the Dilemma

The American militia tradition holds that all able-bodied free adult men have an obligation to bear arms when called to do so by their government. Throughout the colonial and national history of the United States, the militia has filled two roles; it has provided local service, and provided manpower for expeditions against a variety of enemies during wartime. Local service consisted of resisting enemy attacks, suppressing insurrections, enforcing laws, and responding to natural and man-made disasters. Expeditions drawn from the militia have fought Indians, French, British, Mexicans, other Americans, Spanish, and Filipinos. The National Guard, generally regarded as the modern heir to the militia tradition, has gone overseas to fight Germans, Japanese, Italians, Koreans, and Iraqis.

When militiamen depart their homes on expeditionary missions, they leave their communities unprotected. Colonies and states either have weathered wartime without an organized militia under their control, or they have created new militia units for home service during wartime. The uncertainty over the proper role for the militia during wartime increased dramatically during the twentieth century, when the expansion of government responsibility placed increasing burdens on state governments, and the development of airplanes blurred the distinction between the battle zone and the homeland.

Uncertainty over the proper role of militia during wartime began soon after the settling of the first English colonies in North America. The militia tradition brought by the English colonists to the New World predated the American Revolution by several hundred years.[1] The English settlers in Massachusetts Bay formed militia companies in 1636 because they feared attacks by Indians. Colonial militiamen defended the colonies and participated in expeditions against Indians and the French until the War of Independence. During that war, militia augmented Washington's Continental Army, as well as enforced revolutionary discipline among the populace, clearly demonstrating the dual roles of militia during wartime of fighting the enemy and in stabilizing the homefront.[2]

After the end of the Revolutionary War, state militia provided the new nation with its only substantial military force.[3] In part, the desire of the American people to rely on the militia resulted from their fear that a standing army would be expensive and dangerous to the survival of the new republic. A large militia offered a possible solution to the need for a military force that would be competent in wartime yet not dangerous to liberty in peacetime. During the Revolutionary War, the militia had augmented the longer-serving, better-trained, Continental Army, but the Articles of Confederation, in operation from March 1, 1781, until the Constitution went into effect on March 4, 1789, did not provide for the retention or re-creation of the Continental Army after the war. In the concept outlined in the Articles, a "federal" army existed only when regiments raised by the states were called into service for war, revolt, or other emergency. During peacetime, most free adult men belonged to the militia of their state.

The framers of the Articles of Confederation drafted their plan of defense during the war with Britain. They planned a military system to defend the nation against a conventional adversary—Great Britain, Spain, or France. However, the early years of independence presented a very different problem—a seemingly endless struggle with the Indians on the frontier. The idea of a national army of state regiments that existed only during wartime soon proved impractical. The part-time nature of state militia rendered them unsuited for the mission of garrisoning the Northwest Territory. Militiamen with families and farms or businesses could not be expected to remain at frontier posts more than a few months, and the constant rotation of militia units would burden the commanders of the forts by constantly changing their soldiers.

With the adoption of the new Constitution in 1789, the federal government received specific permission to create a federal standing army that was not dependent on regiments organized and equipped by the states. The Constitution recognized the militia in Articles I and II and in the Second Amendment. The martial enthusiasm needed for an effective militia often waned, however, at the local, state, and federal levels. The *Militia Act of 1792* required most free white males between the ages of eighteen and forty-five to arm themselves and attend regular muster, but neither the federal government nor the states enforced the law.[4] The militia, as outlined by the *Militia Act of 1792*, never became the viable nation-in-arms foreseen by its supporters.

Despite the almost immediate breakdown of the militia system in the United States, the American people did not demand, nor did Congress take the initiative to pass, any changes in the national militia law. This failure can be explained by the lack of any immediate or large-scale threat to the nation. The War of 1812 amply demonstrated the weaknesses of America's defense system, yet the results also showed the strengths of the decentralization of the United States. The capture and destruction of the capital by Britain did not bring victory to that country. In addition, the army and the militia occasionally proved adequate for defense, as in September, 1814, when they prevented a British force from burning Baltimore. Moreover, the refusal of Vermont and New York militiamen to enter Canada as part of Maj. Gen. Henry Dearborn's attack on Montreal in November, 1812, on constitutional grounds reinforced the idea that a citizens' militia prevented adventurism and ensured that the military could be used only for defense. The Regulars, however, would not forget their difficulties with the militia.

In the century following the adoption of the Constitution, Congress made no substantial changes to the *Militia Act of 1792*. However, in the years following the War of 1812, the militia as an institution fell into disuse.[5] Few Americans, including Congressmen, saw any need for citizens to waste time drilling when no danger threatened and more profitable pursuits beckoned.[6] A related concept—the Volunteers—became the method by which the states participated in national war efforts.

Volunteers, first authorized in 1806, rested on dubious constitutional ground but solid historical ground. In essence, the Volunteer concept was a holdover from the colonial period, being similar to the means by which the colonies augmented the British Army in the last French and Indian War, and Washington's Continental Army during the War for Independence. While Congress has always had broad powers to "raise and support armies," the Volunteers tended to be companies and regiments raised at the local level.[7] The men from each company elected their officers; the governor of the state appointed the regimental officers; and the regiment was then mustered into federal service for a prior agreed-upon period.[8] Although the authority of governors to raise regiments and commission officers came from the sections of the Constitution covering the militia, the Volunteers were not technically militia while in federal service. Instead, the Volunteers were usually forces raised by the states for a specific period of federal service during wartime. Due to the provisions of the *Militia Act of 1792*, which classed all free adult men as members of their state's militia, the Volunteers were officially on temporary detachment from the militia when they entered federal service. Despite the limitations in the system, it served the government and nation adequately throughout the nineteenth century.

What was commonly called "militia" in the nineteenth century was in reality voluntary military organizations on the local level. In theory all the men in these organizations belonged to the "militia" by virtue of their age, sex, and race, but they were a subset of the militia. A self-selected group, these men formed or joined companies out of patriotism, from fear of slave uprisings, for the comradery, or as a way of establishing social and political contacts, but not out of legal obligation.[9] Therein began a legal fiction: the states legitimized and supported these organizations, and in return these organizations performed traditional militia functions for the state. With these voluntary companies to assist state governments in strike breaking, riot control, and disaster relief, the states were relieved of creating and enforcing a state militia law that followed the *Militia Act of 1792*. The federal government never enforced or changed the Act, and the states ignored it.

The transformation of these so-called militia units into the modern National Guard took the better part of a century. The first use of the term "National Guard" for these organizations came in 1824 when certain units from New York adopted the title as a mark of respect for the Marquis de Lafayette on the occasion of his return to the United States; Lafayette had commanded the Garde National during the French Revolution.[10] By the end of the century, few states and territories had not adopted the term for their militia units.[11] Whatever the name, these state-based companies did not enter federal service during the Mexican-American, Civil, and Spanish-American wars. Instead, they regrouped, often with mostly the same members, as units of state Volunteers that were mustered into federal service.[12] Additional units of Volunteers were raised during wartime from men having little or no militia experience, while members of militia units who did not join the Volunteers remained in their state militia. During the Civil War, many northern states created home guard units that served only in their immediate area during crises, with Missouri's militia system the most complete.[13] Occasionally, home guard units watched prisoners or performed other rear-area duties, thus relieving federal soldiers for combat, as happened during the bloody fighting in late 1864 in northern Virginia. This vaguely defined system for state participation in national defense would last into the twentieth century.

The Regular Army, dwarfed during the Civil War by the Volunteers, remained intact, although Regular officers often took leaves of absence to accept state commissions with the Volunteers. The decades following the Civil War witnessed a resurgence of interest in organized militia. Many Americans believed that the great casualties suffered by the Volunteers in the Civil War reflected the relative inexperience of the newly raised and untrained units of Volunteer soldiers, and more particularly that of Volunteer officers. Men died needlessly until officers learned the trade of war. Advocates of the militia saw the solution in organized companies of militia that trained regularly in peacetime and could then augment the Regular Army in war.[14] In the period after the Civil War, local militia groups, usually including many veterans and men who wished they were, began to take militia training seriously.[15] Although usually still called militia in most states, these organizations constituted the real beginning of the modern National Guard.

States began to take an avid interest in their militia after the violence of the strikes of 1877 and the inability of the militia units then in existence to deal with them. A wildcat strike in late July in the railroad industry in West Virginia over repeated pay cuts soon spread over the eastern half of the nation and into the coal industry. Militia units were either not in existence, sympathetic to the strikers, or not competent for their assigned mission of ending labor violence. In desperation, governors called for federal troops to restore order. Americans realized that class strife was not only a European problem, and middle-class citizens and state leaders took a renewed interest in the militia.[16] From these state and federal concerns the modern National Guard arose.

At the start of the twentieth century, partisans of the militia began to institutionalize the National Guard as a solution to the age-old problem of how to create a military force that would be competent in wartime yet not dangerous to liberty in peacetime. A trained and disciplined National Guard could augment the army during wartime, and during peacetime would allow state governments to maintain order during strikes without the need to call for federal troops.

The growing sense of professionalism within the Regular Army led to an internal debate over the future composition of the wartime army, which arrived at different solutions from the militia proponents.[17] The Regulars concentrated their efforts on divorcing the states from the federal military establishment, giving the federal government a monopoly in fighting land wars. The Regulars desired a reserve force that was wholly controlled by the federal government. They sought to bring the American army into the modern age and saw state military forces as unprofessional holdovers from an earlier era. Their internal debate focused on whether the Regular Army should function as a cadre force that would be able to fight a war only after absorbing recruits to fill its ranks, or remain the same size and create another force during wartime to augment it. Proponents of both views within the Regular Army agreed that the military value of the militia remained negligible as long as states retained control over any aspects of it.

The *Militia Act of 1792* had been an attempt by the First Congress to provide a credible force to augment the Regular Army during wartime without bankrupting the country or resorting to a militarization of American society. These two themes—credibility without militarization—became the focus of the preparedness debate of the early twentieth century. In retrospect, the Spanish-American War represented the swan song of the old system, and the Great War marked the dawn of the modern American military establishment.

Following the Spanish-American War, military planners realized that the old system needed change. However, with Theodore Roosevelt and his Volunteer "Rough Riders" receiving popular credit for the defeat of Spain, the American public did not want to hear about the Regular Army units that cleared the way for his famous charge. Americans retained their traditional disdain for Regulars and continued to champion the amateur soldiers. The concept of the wartime Volunteer as the real military strength of the republic remained entrenched in popular imagination, if not in fact. However, in the decade and a half after the end of the Spanish-American War, the American militia evolved into a very different organization from a legal as well as a military standpoint.

Given the need for organized units and training, the resurgent organized militia—or as it was increasingly called, the National Guard—began to wrest from the Volunteers the official role as the nation's second line of defense. Unlike Volunteers, National Guard units trained during peacetime. Moreover, unlike partisans of the Volunteers, advocates of the militia had a strong constitutional argument. Article I, section 10, of the Constitution stated that "no state shall, without the consent of Congress . . . keep troops," but the oft-quoted Second Amendment guaranteed the right of the states to keep "a well regulated militia." However, Article II, section 3 established the president as the commander-in-chief of the militia when in federal service. Clearly, the framers intended that the militia would be called into federal service when needed, whereas the Constitution made no such direct provisions for states to organize Volunteer units for federal service.

The increasingly professional Regular Army had little but contempt for state military forces. Many reformers in the army were disciples of the late Emory Upton, a Civil War hero and protégé of Lt. Gen. William T. Sherman, who argued that state control of and influence on the militia would always make it unreliable as a reserve for the federal army.[18] Upton used his infatuation with the German military as the lens through which he interpreted American military history. To solve what he saw as the meddling of state politicians in military matters, Upton advocated a reserve force wholly under federal control. The organized militia, even if called "National Guard," was still fundamentally a state force and therefore worthless to the Regular Army. Regardless of the geographical, cultural, and political differences between Germany and the United States, the concept of locally recruited and locally supported units became anathema to Upton's disciples. Instead, they sought total federal control over recruiting, organizing, and leading of any non-Regular forces to be employed in war.

On the opposite side of the argument stood a powerful lobbying group in Washington that opposed the creation of a federally controlled reserve. The National Guard Association (NGA), a nationwide organization of National Guard officers formed in the late nineteenth century, lobbied for recognition of the National Guard and against the creation of some new federal military reserve as the second line of defense behind the Regular Army. State governments supported the National Guard because of its potential service to the states during riots, floods, and strikes, but most National Guardsmen themselves eyed a federal role during wartime.

The National Guard of the late nineteenth century, although better organized, equipped, and trained than most previous incarnations of the American militia, was still an awkward vehicle by which to expand the army in wartime. The National Guard, despite its name and its origins in the resurgences of nationalism in the late nineteenth century, bore a heavy imprint of state identity among its units. The National Guard was not standardized across the nation or with the Regular Army. Each state, territory, and the District of Columbia had its own uniforms, training programs, and organization. To integrate this polymorphous force into federal service would be an extremely difficult task.

The *Militia Act of 1903*, called "the Dick Act" after the bill's sponsor, Congressman Charles F. Dick of Ohio, sought to bring the National Guard in line with the Regular Army.[19] The former federal law for the militia, *Militia Act of 1792*, had never been followed, and the rise of the National Guard made its inadequacy all too apparent. Congressman Dick served as president of the National Guard Association and was the Commanding General of the Ohio National Guard. He had originally joined the National Guard in 1876 and later served as a lieutenant colonel of Volunteers in 1898. As president of the NGA, Dick had been a strong force in favor of a war-fighting role for the National Guard.[20] The Dick Act recognized in federal law for the first time the distinction between the organized and unorganized militia—which had existed in fact for a century. The old concept of universal adult free male militia obligation inherent in the old *Militia Act of 1792* had never become a reality, but federal law had taken no notice of the distinction between men who actually belonged to militia units and men who had never stood a single muster. The new *Militia Act of 1903* defined the organized militia as those military organizations, primarily the National Guard, which had been organized and recognized by the states.[21] The unorganized militia consisted of the remainder of the male population liable for militia service should the state someday choose to organize it.

The Dick Act gave official recognition to the term "National Guard" for the land forces of organized militia of the states.[22] At the time, most militia units trained one evening a week, although attendance varied widely, and some units seldom drilled at all. The Dick Act authorized federal funds for at least two mandatory drills per month.[23] National Guard units also had to hold a minimum number of target practices every year. The *Militia Act of 1903* required militia units to spend five days training in the field annually.[24] Each state's militia had to follow the Regular Army in organization, equipment, and discipline.[25] The planners hoped the changes would mold the National Guard into an effective reserve force that the federal army could and would use in war. State governments generally supported any federal move that would bring more equipment and money to the National Guard because the states would then have at their disposal a better organized, equipped, and trained militia during peacetime. Neither the states, the army, nor the National Guard seem to have considered the question of what would replace the National Guard in its state mission should it ever be called upon to fulfill its federal mission.

The Dick Act began the transformation of the various groups of militia into the modern National Guard. However, constitutional issues remained unresolved. The National Guard remained legally militia. With the National Guard organized under the militia clauses of the Constitution, it could not be used outside the borders of the United States. In addition to constitutional limitations, *Militia Act of 1903* contained many clauses that left the National Guard poorly positioned to augment the Regular Army in war. State governors retained the right to authorize or to deny their state's units to enter federal service. Even with the governor's consent, individual National Guardsmen needed to volunteer for federal service; they could not be drafted against their will.[26] In addition, federal service could not extend beyond nine months in a year.[27] Finally, no federal agency, including the army, had the authority to remove militia officers, even while in federal service, no matter how incompetent they were.[28] The limitations made the National Guard little better than the Volunteers. Amendments to the *Militia Act* in 1908 lifted the nine-month time limit and gave the president the right to prescribe the length of time militia units could spend in federal service, although individual National Guardsmen could not be forced to serve past their existing enlistments or commissions.[29]

The changes of 1908 further provided that the National Guard could be ordered for service "either within or without the Territory of the United States."[30] However, in 1912, the attorney general ruled that sending National Guard units outside of United States territory violated the Constitution.[31] In his decision, he reasoned that the National Guard was militia, and as such it could only be used for the constitutionally mandated functions of suppressing insurrection, enforcing law, and repelling invasion. The militia could only leave United States territory in pursuit of an invading army. The opinion of the attorney general that the National Guard could not be used beyond the borders of the United States delivered a seemingly mortal blow to the National Guard's desire to be the nation's second line of defense. Fearing that the National Guard might be shunted aside in favor of a new federal reserve, the National Guard Association lobbied hard for a solution to the constitutional barrier that would allow the National Guard to serve beside the Regular Army anywhere in the world.[32]

In a 1914 amendment to the Dick Act, the National Guard received the official designation as the nation's second line of defense, but the lack of a mechanism for overcoming the constitutional barrier against militia leaving the United States weakened this victory. National Guardsmen could enter federal service as a company only if three-quarters of their unit volunteered for federal service. In wartime, the president had to accept all National Guard units that met this criteria before he could accept any units of Volunteers.[33] In essence, this change meant that the National Guard could fill the role of the Volunteers without losing their lineage and formations, but little else had been achieved.

In the preparedness debate which began in America after the summer of 1914, the abstract speculations of the 1880s became urgent arguments as the nation sought the proper balance of trained professional soldiers and patriotic citizen-soldiers in a world that suddenly seemed far more dangerous. Secretary of War Lindley M. Garrison developed a plan for a reserve force to replace the National Guard as the country's main augmentation of the Regular Army. Convinced that state control of the National Guard would always hamper its usefulness, Garrison planned a federal military reserve that he called the Continental Army, in a shrewd move to link it with patriotism and the Founding Fathers.[34] Garrison's full program of military preparedness called for an expansion of the Regular Army, increases in the navy, and a 400,000-man reserve force completely under federal control.[35] Members of the Continental Army were to be civilian volunteers who trained under army supervision during the summer. Worried about the war in Europe, Pres. Woodrow Wilson, who had shifted from his traditional opposition to the military to one in favor of preparedness, backed Garrison's program. Although he reassured the National Guard that he had no plans to dissolve it, the President clearly had been influenced by the arguments of the disciples of Upton.[36]

The secretary's plan met with bitter resistance from the National Guard Association, which correctly feared it was an attempt to shuffle the National Guard into obscurity. Such a plan would transform the National Guard, which had formed primarily to allow competent state participation in national war efforts, into a permanent home guard force, to serve only within their states. To placate the National Guard, it would receive a slight increase in federal support but would be relegated to local militia duties in war and peace. However, traditional suspicion of military expansion remained strong, especially in the president's own party, the Democrats. The secretary of state, William Jennings Bryan, broke with the administration to fight against the plan. Bryan felt, as did many opponents of preparedness, that Americans would fight if the nation were invaded. However, as no nation seemed about to invade, preparedness would be the first step on the road toward European-style militarism.[37] Many southern congressmen opposed the plan in part because they feared that blacks, who were excluded from the National Guard in the South, would join a federal reserve force in large numbers.[38] When Garrison testified before the House committee, he admitted that many of the men for the Continental Army would most likely come from the National Guard, but as the National Guard and Regular Army could not recruit to strength, the government would probably need to draft citizens to fill the proposed reserve of the army. That revelation killed the plan for the immediate future because the country as a whole was unwilling to support mandatory military service. In the years before the Great War, Americans found conscription as repugnant as had their grandfathers during the Civil War.[39] Americans opposed forcing men into uniform before an actual state of war existed.

Americans' opposition to conscription did not preclude allowing men to train voluntarily under army supervision during peacetime. Reformers who believed in the concept of the patriotic citizen who leaves civilian pursuits for military service during wartime hoped that two programs would provide trained officers for the Volunteers in future wars. Beginning in the summer of 1913, the army had conducted a series of Student Military Training Camps. Students from the nation's colleges and universities who signed up for the camps spent four or more weeks learning basic military skills. (The number of weeks varied depending on the year.) Each participant paid for his uniform and meals. After completing the course, his training with the army finished, each "veteran" was expected to spread the gospel of the need for preparedness and the benefits of military training.[40] Related to the summer camps was military instruction at land-grant colleges. The *Morrill Act* of 1863 required that all land-grant colleges teach courses in military tactics.[41] Some, such as Norwich University in Vermont and the Virginia Military Institute, conducted highly credible programs. At the majority of schools, however, the training was considered a boring joke by the students and the equivalent of exile by the officer-instructors. Moreover, the War Department kept no record of students with military training once these students had graduated from college.[42] Maj. Gen. Leonard Wood, the army chief of staff who had earned his commission through the Medical Corps and not West Point, began to agitate for cooperation between the War Department and the Interior Department to standardize military instruction at land-grant colleges. He hoped to give provisional commissions, what today would be called Reserve Commissions, to honor graduates, who, after a year of active training in the Regular Army, would form an officers' reserve, which would provide the leaders for the Volunteers. That the Volunteers would still be needed in future wars was taken for granted.

As events were to show, America's first line of defense, the Regular Army, was stretched thin by the increasing burdens placed upon it. America's flirtation with imperialism at the turn of the century left the army with additional missions far removed from the continental United States. In addition to manning coastal defenses and posts in the West remaining from the frontier era, the army maintained a large minority of its strength in Hawaii, Alaska, the Canal Zone, Puerto Rico, and the Philippines. With Europe at war and direct United States involvement looming, the whole military establishment found itself overtaxed responding to a small raid on the country's southern border. In early March, 1916, Francisco "Pancho" Villa raided the town of Columbus, New Mexico, with a force of about five hundred men. The army sent a force under Brig. Gen. John J. Pershing into northern Mexico with the mission of destroying Villa's army. A raid two months later on two Texas border towns by bandit gangs from Mexico led President Wilson to call the National Guard of every state into federal service to protect the border. Under *Militia Act of 1903*, the status of the National Guard remained unclear. Although it was the main reserve of the army, legally it remained militia and was therefore limited to service in the territorial United States—it could not cross national borders. No solution to the constitutional limits of a state-based second line of defense had yet been found. The restriction led to such farces as commanders allowing the cavalry men of the Virginia National Guard to ride their horses up to the Mexican bank of the Rio Grande by the town of Matamoros, but forbidding them to leave the river on the Mexican side.[43]

As the bulk of the Regular Army and National Guard sweated in the heat of the American Southwest and northern Mexico, Congress began to create a new law that would better allow the army to respond to the changes in its mission since the era of the Founding Fathers. The final product, the *National Defense Act of 1916*, bore the imprint of the preparedness debate, the shortcomings exposed by the army's response to Villa's raids, the lobbying power of the National Guard Association, and the ghost of Emory Upton. Once the *National Defense Act of 1916* became law, the relationship between the Regular Army and the National Guard fundamentally changed. The *National Defense Act of 1916* would provide the basic legal framework of the United States military until after World War II. Through it, Congress tried to correct many of the shortcomings of the militia left unsolved by the *Militia Act of 1903*. Spillover from the Mexican Revolution, the preparedness debate, and the war in Europe all served to convince Americans to abandon the archaic model of national defense inherited from the revolutionary generation.

The *National Defense Act of 1916* altered the constitutional status of the National Guard when called into federal service. The new law defined the Army of the United States as consisting of the Regular Army, the Volunteer Army, the Officers' Reserve Corps, the Enlisted Reserve Corps, and the National Guard when in federal service. In truth, only the Regular Army and the National Guard actually existed.[44] The president could augment the Regular Army with the National Guard for the duration of a national emergency in accord with the changes of 1908, referring to its use for more traditional militia functions as in repelling invasion or patrolling the Mexican border. Other sections of the *National Defense Act of 1916* would allow the National Guard to accompany the army beyond the national borders. The constitutional objections against sending the National Guard outside of the country were solved by requiring all its officers and men to take a new oath swearing to defend the United States as well as their home state. When the National Guard was brought into the Army of the United States for war, all National Guardsmen were discharged from the National Guard and became soldiers in the federal army, although they would keep their same unit designations.[45]

"Section 57" of the Act reasserted the right of the federal government to conscript men into military service. President Wilson claimed that when the Republic declared war, its citizens had in effect all volunteered for military service.[46] The new law put this idea into legal form. "All able bodied male citizens of the United States" and those intending to become citizens, between the ages of eighteen and forty-five, were declared to comprise the militia of the United States. Men not in the Regular Army, or in their state's organized militia—the National Guard or naval militia—formed the so-called unorganized militia.[47] This formulation meant that as residents of a democracy whose elected officials declared either a war or other emergency they were eligible for military service. By electing the senators who had declared war, the people had in effect volunteered for active military service to fight that war. It also meant that free adult male civilians who were not in the National Guard or naval militia belonged in theory to two separate unorganized militias—that of their state and that of the nation. Either government could conscript and organize them, although federal conscription took priority.

Another section standardized the number of National Guardsmen in each state. Within one year of passage of the act, each state was to have two hundred enlisted men for each senator and representative that the state sent to Congress.[48] After the first year, the plan called for the size of the National Guard to increase by 50 percent each year until the number of enlisted Guardsmen reached at least eight hundred per congressman.[49] As National Guard units were already under-strength, the possibility of recruiting to higher strengths was not likely to happen in peacetime. However, with the United States formally at peace, Congress hoped to provide a reserve force without resorting to the conscription implicit in the plan for the Continental Army.

The new law also better integrated supervision of the National Guard by the War Department. The army had created a National Militia Board to advise it on militia matters in 1908.[50] The new law moved militia affairs up to bureau level, although the chief of this new Militia Bureau would be a Regular Army officer and not a National Guard officer.[51] Recognizing the need for closer cooperation between the Regular Army and the National Guard during training, the new law provided for 822 extra officers and 100 extra sergeants from the combat arms of the Regular Army to serve one-year tours as instructors with the National Guard.[52] In a move that allowed the War Department better to integrate the National Guard into its war plans, the president received the authority to decide which types of units states were to maintain; prior to the *National Defense Act of 1916*, National Guard companies adopted whatever branch they fancied. After the *National Defense Act of 1916*, the president, through the army, would decide whether a National Guard unit would be cavalry, artillery, quartermaster corps, or whatever the army needed.[53]

The *National Defense Act of 1916* became the watershed that began the modern National Guard. Under it, officers in the National Guard were to apply for a federal commission with the National Guard of the United States as well as the National Guard of their state, while enlisted men were to enlist in both the National Guard of their state as well as the reserve of the army. When inducted with their units into the Army of the United States, National Guardsmen were, in effect, discharged from the organized militia of their state and brought into the federal army as part of the National Guard of the United States. The National Guard when brought into the Army of the United States ceased to be organized militia and was thus unhampered by the constitutional limits on the use of militia. Once inducted, National Guardsmen fell under the sections of the Constitution that gave the federal government the power to "raise and support armies."[54]

The evolution of the National Guard into a reserve force for the federal army achieved most of the goals of the National Guard Association. National Guardsmen received equipment and pay from the federal government as well as a guarantee of a wartime role as units in the federal army. But if the National Guard Association had worked long and hard to prevent a federal monopoly of military force, it was less generous regarding other state military forces. The *National Defense Act of 1916* contained what appeared to be a clear policy on other militia forces:

> Sec. 61. Maintenance of other troops by the states. No state shall maintain troops in time of peace other than as authorized in accordance with the organization prescribed under this act: Provided, that nothing contained in this Act shall be construed as limiting the rights of the states and territories in the use of the National Guard within their respective borders in time of peace: Provided further, that nothing contained in this Act shall prevent the organization and maintenance of state police or constabulary.

Although opinions differed, most states saw the section as prohibiting the formation of an organized militia that was not part of the National Guard or naval militia.[55] State police, unlike militia, were full-time and operated under civil laws. "Sec. 61" would cause confusion and uncertainty within a few months of the adoption of the Act.

Americans in 1916, no less than their predecessors in 1789, tended to distrust a large standing army and to place great faith in the martial abilities of the civilians who took up arms when war began. The *National Defense Act of 1916* was a compromise between the various schools of thought on preparedness. The Volunteer system remained extant in theory, with an expansion of ROTC and Plattsburg-type camps to train a reserve of its future officers.[56] But the Volunteers were relegated to the third tier of the nation's defense, after the Regular Army and National Guard. The *National Defense Act of 1916* guaranteed that in the next war, the entire National Guard would be taken into federal service before any additional forces were accepted. The National Guard Association got most of what it wanted in the Act, and members took the dual oath to became the nation's official second line of defense. Through the National Guard, the states maintained a vehicle to participate in future wars. The National Guard remained a state-based military force, with federal pay and equipment, but with the ability to accompany the Regular Army into any theater of operation in the world. Only one issue remained undiscussed, unsolved, and perhaps unanticipated: The *National Defense Act of 1916* ensured that states would lose their National Guard—their organized militia—during America's next war, and no thought had been given to what would replace it for state duties. And that next war was already under way in Europe.

CHAPTER 2
Home Guards of the Great War

During the Great War, states lost their National Guard units when they entered federal service. The various responses of the states demonstrated the advantages of maintaining "a well regulated militia" and the perils of not. The induction of the National Guard into the army in 1917 had been the realization of a long-sought goal of military reformers. The old Volunteer system had been replaced by National Guard units that trained during peacetime for service in war; however, a basic contradiction inherent in the dual role of the National Guard had never been addressed. The reforms in the *Militia Act of 1903* and the *National Defense Act of 1916* had simply not mentioned the issue of state service while the National Guard was in federal service. "Section 61" of the *National Defense Act of 1916* granted recognition to the National Guard as the only organized militia, while other sections of the act mandated that, in any action requiring more military power than the Regular Army could provide, the president call the entire National Guard into federal service in advance of any other forces to be raised for war. When President Woodrow Wilson drafted the entire National Guard into the army in the summer of 1917, the contradiction between the state and federal missions of the National Guard came to a head for the first time.

A few days before the declaration of war by the United States in April, 1917, President Wilson brought most of the National Guard onto federal active duty to patrol transportation, industrial, and communication centers in each state. Although on active federal service, the National Guard in effect was still organized militia, albeit under federal control. The Constitution gives the president ample authority to call the militia into federal service. Although when that happens the president becomes commander-in-chief of the militia, it remains separate from the army.[1] The National Guard remained the organized militia of the states in federal service. As such most units remained in their home state, maintained their state structure, and could not leave the territory of the United States. The National Guard entered federal control under the militia clauses of the Constitution but would be mustered into the United States Army as per the *National Defense Act of 1916* during the summer of 1917.[2] By law, the president had to induct the entire National Guard into the army before raising any additional land forces.

On June 2, 1917, the army sent a circular letter to all state adjutants general forwarding orders from Maj. Gen. Leonard Wood. The letter explained that because soldiers wore uniforms while protecting industrial sites, they were useless for uncovering sabotage plots. Instead, the army suggested the use of civilian watchmen to protect factories, retaining military personnel in that capacity only for protecting "certain railroad bridges and tunnels, the destruction of which would cause prolonged delay in the service between important centers and the guarding of which cannot with safety be performed by civilian watchmen."[3] The National Guard was to be absorbed by the army starting June 15, and all National Guardsmen were to be withdrawn from local duties by August 5. The only exceptions were a few places deemed of vital national importance, the destruction of which would seriously affect the war effort, such as key railroad bridges and tunnels.[4]

When the president drafted the National Guard into the army under the provisions of the *National Defense Act of 1916*, he severed all its remaining connections to the states. States no longer had any control over their National Guard units in federal service, and individual National Guardsmen had no obligation to the states. The induction of the National Guard left states during wartime without their principal means of responding to natural disasters, suppressing riots, and assisting local lawmen when mobs threatened to lynch suspects in custody.[5] Only Pennsylvania had a state police force, and most states depended on their National Guard for police functions. To replace the National Guard in its state militia role, many states permitted the creation of loosely supervised units of home guards, while other states created more centralized military forces. The most credible of these forces tended to come from the wealthier and more urbanized states in the Northeast. Massachusetts and Connecticut, in particular, created forces similar to their departed National Guard in most respects, but maintained completely with state funds and without any federal obligations. By contrast, most southern and western states took a decentralized approach to replacing their National Guard for state service. As a result, states with a true organized militia spent the war better equipped to meet wartime emergencies and were able to respond to the unrest that swept the nation in the year and a half following the Armistice.

Massachusetts and Connecticut created effective and centralized state military forces that they christened the "State Guard" to stress the statewide obligation of these forces. The State Guards of Massachusetts and Connecticut resembled the departed National Guard, although because of federal conscription, State Guard units had an enormous annual turnover of membership, and State Guardsmen tended to be either older or younger than their National Guard counterparts. State Guardsmen provided needed services to their states during disasters, the "Spanish Influenza" epidemic of 1918, and periods of civil unrest. By contrast, in the poorer and more rural Deep South, governors and state adjutants general reacted with indifference or hostility to locally created home guards. Lacking uniforms, support, or encouragement, southern home guardsmen contributed little service to their states.

The difference in response of the two regions reflected differences in wealth as well as history. The more industrialized states of the Northeast could better afford to create, equip, and uniform a new state-supported militia. In the years before the entry of the United States into the Great War, the National Guard had become increasingly dependent on federal dollars. Under the lobbying efforts of the National Guard Association, the National Guard had evolved from a state militia that the federal government could use during a national emergency into a reserve of the army that states could use for militia functions when it was available. Although states provided a small part of the funds and built some of the armories for the National Guard, the federal government provided the bulk of pay, equipment, and weapons for it. New militia raised as a wartime replacement for the National Guard had no federal mission and would receive little federal support. Southern states had fewer resources and could not afford to create forces similar to the National Guard. In addition, the excesses of home guards during the Civil War gave southerners reason to fear a loosely controlled militia as much as potential unrest. Local groups called home guards often terrorized the countryside as a combination of vigilante and private army bent on survival.[6]

In 1917, although southerners still feared loosely controlled forces, most southern state governments did not create a new, centralized militia. Prior to the American entry into the Great War, Florida had two regiments of National Guard. In June, 1916, following Villa's raid on Columbus, New Mexico, the Second Florida Regiment of Infantry mobilized for service on the Mexican border of Texas. To bring the regiment to full strength, volunteers from the First Regiment joined the Second. After the Second Regiment left Florida, the much diminished First Regiment, redesignated the First Separate Battalion, remained available for state service. As such, it performed active service guarding bridges. In April, 1917, it saw service protecting a suspect from a lynch mob in Hillsborough County. After a few tense moments, the mob dispersed without testing the resolve of the Florida National Guardsmen.[7]

Shortly after this incident, both of Florida's regiments entered federal service for the Great War. Following their draft into the army, the First Regiment was divided between new units, and the Second Regiment formed part of the Thirty-first "Dixie" Division.[8] The departure of Florida's two National Guard regiments of infantry for the war left the state with no remaining military forces.[9] The state legislature on May 21, 1917, passed a law allowing counties, through their Board of County Commissioners, to form County Guard units composed of white men between eighteen and sixty-five years of age.[10] Each unit received a charter as a private corporation from the state. County Guard units were to be dissolved within four months after the end of the war. Sheriffs and judges had the authority to order out the force for peacekeeping.[11] Florida's governor, Sidney Catts, Sr., managed to acquire 750 federally owned rifles, but because the units were county forces and he was a state official, he did not have the authority to sign the bonds that the federal government required before issuing federal property. Without a way to issue the weapons legally, he returned them to the federal government. He then tried unsuccessfully to get the state legislature to put the County Guards directly under his control.[12] Apparently Florida's legislators desired to keep the new militia a purely local force and not have the state assume the financial burden for maintaining it.

Only two instances of active duty for the Florida County Guard units are recorded. In October, 1917, county authorities in Madison County, on the Georgia border, called out the Duval Guards to prevent a possible lynching of a suspect in a criminal trial. This action involved a single company, composed mostly of professional men, or men too old for federal service. The state legislature eventually paid the men for their five days of service, even though they were acting under county authority.[13] The other incident came in August, 1919, when a single officer was sent to mediate during a strike in the phosphate industry in Mulberry, thirty miles east of Tampa. Apparently he did his job well, because the strike ended without the need to call for more Guardsmen.[14] Despite these isolated successes of the County Guards, Adj. Gen. Sidney Catts, Jr., the governor's son, said that his experience with the County Guards was "far from satisfactory" and that their main use had been in parades and ceremonies. His father, shortly after the end of the war, told the legislature that "the Home Guard system failed to achieve its purpose," without elaborating what he believed that purpose to have been.[15] Like many other states without a wholly state-funded and-controlled force, Florida found little to celebrate in its wartime militia.

Even less successful than the County Guards of Florida was Alabama's response to the loss of its National Guard. When the Alabama National Guard left the state to enter the army, Alabama also responded with a decentralized approach. The state government created no organized militia but instead encouraged—although little more—sheriffs throughout the state to enroll local men for emergency duties. On June 5, 1917, Governor Charles Henderson sent every sheriff in Alabama a letter explaining that with the National Guard in federal service, getting federal troops for local emergencies would be a time-consuming process, and with the induction of the National Guard into the army imminent, National Guard troops would soon be completely unavailable to sheriffs. He therefore suggested, although he did not require, that each sheriff use his own authority as executive officer of each county to summon together about fifty volunteers of responsible men, to be called the Home Defense Guard. Each unit was to operate only within their own county, under the authority and direction of the sheriff. In practice, larger towns created companies of fifty men or more, while smaller communities usually enrolled a two- or three-man "beat"—men who agreed to respond to call from their sheriff—or nothing at all.[16] Although cost-free to the state treasury, this stop-gap measure was never much more than a plan.[17] Most of the units existed only on paper, although some units became more substantial through local initiative.

An example of one of the more substantial units existed in Albany, although it did not meet with the approval of the governor. Mr. B. L. Malone, a self-styled major who headed the city of Albany's militia, wrote several letters to the governor, and later, the War Department, in his quest to acquire arms for his men. After the War Department informed Malone that creating militia was a state function, he again wrote to the governor in exasperation pointing out a detail he felt sure the governor must have overlooked: "Special Regulation Number 37," one of Alabama's wartime ordinances, explicitly gave governors the authority to issue weapons from the War Department to home guards. He explained to his reluctant governor that if the War Department was in favor of arming home guards, then it must have approved of their creation. The letterhead on the major's repeated requests contained the claim that the force was "armed by the federal government and subject to the call of the governor,"[18] but in reality, the federal government never armed the force, and the governor never called on it. Another militia formed in Talladega, but soon disbanded after the local sheriff opposed the group as undermining his authority.[19]

Throughout the war, the governor was swamped with requests from citizens of the state asking for special consideration for commissions in regiments of state volunteers for federal service, which would never be raised. Most people in the nation did not realize it, but a fundamental change had occurred in the way the United States went to war. The old system whereby men who wanted to participate in the war formed militia companies that were then detached from their state militia and accepted into federal service for a limited period of time had ended. Although the *National Defense Act of 1916* did authorize President Wilson to accept four divisions of Volunteers, he made no plans to do so, mostly out of fear that his political rival Theodore Roosevelt would end up leading one.[20] For the Great War, the National Guard and Selective Service became the vehicles through which the small standing army expanded for war. The partisans of military reform had achieved most of what they had sought after the Civil War. Militiamen had trained in peacetime through their National Guard before they entered federal service as units for war.

Large numbers of men assumed that war meant volunteer regiments would again be created by the states for federal service. The fundamental changes wrought by the rise of the National Guard had not reached the consciousness of society. Many young men eager for glory on distant battlefields assumed that their path took them through their state governor. Governors also heard from large numbers of citizens who considered themselves unfit for the arduous business of fighting a war but still wanted an official role, such as an appointment to selection boards. In addition to these requests for roles related to the national war effort, many men, and a few private organizations, petitioned the governor for the right to form militia companies or to seek commissions in the home guard they assumed would be created, but the governor held firm in his county sheriff plan. One such request came from a man in Hackleburg, a small town ten miles south of Muscle Shoals, who required the services of a local merchant to write his letter. The letter explained that, although the man was registered with Selective Service, he would most likely be deferred because he had a wife and children. Still, he wanted to do his part, and sought an appointment in the home guard, which he believed the governor intended to form.[21]

One of the private organizations seeking a public role during the war was the "Uniform Rank, Woodmen of the World," which wrote to the governor on July 19, 1917, explaining that the organization had fifteen companies of thirty-two men each stretching from East Lake to Bessemer, near Birmingham. The parent organization had given the Alabama chapters permission to "change its arms from Axes to Rifels [*sic*] 'for drill and parade purposes.'"[22] They sought the governor's permission to begin their wartime conversion to militia. The governor replied to this group that several other organizations had made similar requests, but he had denied all of them because he disliked the idea of allowing the existence of armed organizations not directly under civil authority. He reiterated that he intended to encourage local sheriffs to arrange for emergency forces.[23] Perhaps part of his opposition to loosely controlled militia forces came from a fear that they would become in reality vigilante groups. One small-town merchant wanted to form a home guard to "catch German Spies and any one that might be opposed to the Government OFThe [*sic*] United States." He offered gratis the services of this militia he planned to raise.[24] Despite the enthusiasm of this petitioner, the governor wanted no such forces with vague and potentially dangerous missions loose in his state. For practical purposes, Alabama weathered the Great War without an organized militia at the state level. Only in November, 1918, just days before the Armistice, did Alabama's adjutant general, who had played almost no role in the sheriff-based plan, begin the formation of militia companies for home service. This new force was specifically called "National Guard," on the promise of the War Department that although it would arm and equip new units as equipment became available, they would not be taken into federal service.

Next door to Alabama, in Mississippi, the adjutant general, Brig. Gen. Erie C. Scales, took more interest in local home guards, but nevertheless, did not encourage their creation, and fought to disband them immediately after the war. The initial reaction of individuals and communities in Mississippi reflected regional and national concerns in the early days of the involvement of the United States in the war. In the days following the declaration of war, several communities either formed home guard companies, or requested instructions on how to do so, to replace the departing Mississippi National Guard as organized militia.[25] General Scales received the first such request on April 6, 1917, from the postmaster of McComb, W. W. Robertson. Robertson requested permission to form a company of twenty men armed by the government because the town included "a German and Austrian element" who "seem all right, but you can't tell always what a man or woman is by appearance."[26] The following day General Scales composed a reply that would soon become his standard response. After acknowledging the request he explained that the "National Defense Act of June 3, 1916, as well as the Military Laws of Mississippi [made] no provisions for a Home Guard" and that therefore the authority and request had to be denied.[27]

On the same day, R. T. Luke of Laurel, Mississippi, wrote to the general requesting information on the legality of forming a home guard company that either the federal or state government would supply. He specifically stated that "[w]e do not care to be affiliated with the National Guard or Regular Army, either."[28] Again the general replied that the laws did not permit the creation of such companies.

In at least two small towns, the proprietor of a drugstore took the initiative in forming organized militia companies for local protection. Mr. C. E. Anding of Peoples Drug in Leakesville wrote to the general that "[t]he business interests here would like to see this accomplished and will contribute liberally to the support and maintenance of a company." He explained that several people had asked him to provide training for them because he had received some military training while in school.[29] On the same day, W. A. Hickman of Hickman's Drugs of Monticello also requested permission to form a company.[30] The general sent his usual response.

By the end of the month, General Scales had heard from at least eleven individuals or groups requesting information about forming home guards.

To a group of eager young men between the ages of fourteen and eighteen he suggested forming a Boy Scout patrol or raising and guarding food.[31] The state secretary of the Knights of Pythias, a fraternal organization with military trappings, offered to form a "National Guard company" of over-age men for coast protection.[32] A man in Inverness proposed to form a "Guard" company of married men for service inside the United States, preferably in Mississippi.[33] Residents in the town of West Point took the most initiative. Shortly after the declaration of war, the town held a mass meeting where volunteers formed a battalion of three companies, whose members then elected their officers. The battalion had actively drilled and recruited for a week when an assistant to the battalion commander informed the adjutant general of the actions taken by the townspeople and requested manuals of arms. Although the general, in his reply, reiterated that such formations were not sanctioned under state or federal law, he mentioned that Congress had taken up the issue and was working on legislation that would provide for home guard companies. In the meantime, he denied the request: military publications were issued to the state only for use by the National Guard.

Only one request for permission to form an organized militia came from a town official. On April 26, the mayor of Tchula requested information on forming a company of home guards.[34] A more colorful request came from a group of "Ex-Confd Soldiers" from Patterson who wanted to form a company of home guards and wanted to know "how to proceed." The letter was signed, "Yours for Victory, whatever the cost."[35] Again the general advised that, while current law did not allow such companies, Congress was working on legislation that would soon provide for such units.

After receiving a flood of requests either to authorize the formation of home guards, or to recognize locally created units, Scales reluctantly acknowledged six companies of home guards. The necessary legislation had been passed at a special session of the Mississippi legislature that summer. Of the eleven towns that had wanted to form companies in April, 1917, only one, Meridian, was assigned a company. The others were based in Enterprise, Oxford, Scooba, Seminary, and Vicksburg.[36] Only Enterprise and Seminary, towns close to Meridian, did not normally have a National Guard unit based in their town.[37] Four of the six units were located within seventy-five miles of Meridian. Meridian concerned the adjutant general.

Communities feared two basic scenarios. "Pancho" Villa's raid on Columbus, New Mexico, had occurred slightly over a year previously, and the Punitive Expedition had left Mexico only months before America declared war against Germany. The Mississippi National Guard had not returned from the border until February, 1917. Stability appeared to have returned to Mexico, but fears of an alliance between Mexico and Germany renewed anxiety over the security of the border. Rumors of German officers training the troops of Villa were rife throughout the nation.[38]

Perhaps the greatest fear for white Mississippians came from their uncertainty over the attitudes of black Mississippians to the war. Rumors abounded that German spies were fomenting unrest among blacks. Whites in Meridian, on the eastern side of the state, seemed most prone to this fear. Meridian had a large black population, and from the first days of American involvement in the Great War into the early 1920s, racial tensions reached levels higher than any other time since Reconstruction.[39] Apprehension among whites increased when the National Guard from Meridian left town on April 5 to help patrol the levees on the Mississippi River, which had reached flood stage.[40] Across the state in Vicksburg, on the Mississippi River, tensions and violence between blacks and whites exceeded even the level in Meridian.[41]

Even if the threat of Mexican-German invasion or German-inspired race riots had been real, the communities did not yet need home guards. The National Guard remained within the borders of the states. Despite the declaration of war by the United States, the president waited until summer to draft most of the National Guard into the federal army. Instead, National Guard units had been on active federal service since late March, 1917, in order to protect transportation, industrial, and communication centers within each state.[42] This situation could not last indefinitely because the army needed to absorb the National Guard for rapid expansion for the war in Europe.

Shortly after the National Guard was drafted into the army in August, 1917, southern governors received a letter from General Wood that addressed the problem created by the departure of the National Guard.[43] The letter reiterated the basic points of his earlier circular letter. Wood mentioned that "no serious attempts have been made to damage or destroy public or private utilities on our Southern States. The people are loyal and can be depended on."[44] The second page of the letter addressed the problem created by the departing National Guard. Wood acknowledged that legitimate questions of local security remained and that businesses that could not provide their own protection would "natural[ly]

. . . look to State or municipal Governments for aid." As a solution, General Wood quoted "Section 61" of the *National Defense Act of 1916*, with emphasis on the last line, which read that "nothing contained in this Act shall prevent the organization and maintenance of state police or constabulary." Apparently in General Wood's view, this provision allowed states to create a separate organized militia force for the duration of the war. He explained the system in New York, where a "State Constaublary" [*sic*] had been created specifically for the governor to use while the state's National Guard was in federal service. He also mentioned that Pennsylvania had created a constabulary similar to the Northwest Mounted Police of Canada and that other states were organizing home guards to replace the National Guard. Whatever form each state adopted, General Wood assured the governor that the members of the force would have "the authority of the State behind them, for the proper performance of their duties, and to enforce State and Federal Laws."[45]

The day after General Wood wrote his letter, Congress passed a bill that specifically authorized the secretary of war to issue rifles, ammunition, and military kit not needed by the federal forces to the states and territories. Although the bill mentioned that the equipment was to be used by state police or constabulary, it also specified that home guards could receive the equipment. The governor of a state receiving such equipment had to sign a receipt for the equipment and place a bond for it. The equipment would remain the property of the federal government and was subject to recall at any time.[46] But because of the vast expansion of the armed forces, the secretary of war had no extra equipment to spare for state military forces.

While the Mississippi state legislature authorized six companies of home guards that had formed without state direction or help, it neglected to provide funds for arming and equipping them. For fiscal year 1918, the state legislature appropriated $16,849 for the National Guard with a further $15,849 in 1919. As the state's National Guard technically included only those few National Guardsmen, such as General Scales, who had not been drafted into the United States Army, the money was used instead in an attempt to purchase equipment from the civilian market and weapons from the federal government. Owing to the large amounts of supplies needed by the federal military establishment, the state could not purchase any supplies.[47] General Scales believed that, had an emergency arisen, the men in the home guard companies would have been able to arm themselves from their own resources. However, the little force received almost no help from the state. The Mississippi Home Guard of the Great War, unarmed, un-uniformed, and unequipped, inspired little confidence in General Scales, who referred to them as having "little value."[48]

In the spring of 1918, Gov. Theodore G. Bilbo instructed General Scales, in his capacity as commanding officer of the Home Guard of Mississippi, to report to the chief of ordnance, United States Army, on the status of the home guards. In the report, the general indicated that the force did not have uniforms, distinctive insignia, or weapons. He stated that he hoped to receive enough rifles from the federal government to equip a force of 1,000 men, and that if he did not, he would disband the force.[49] Under federal provisions, Mississippi was to receive 1,700 rifles originally intended for sale to Russia, and 55,000 rounds of ammunition. Although the state filed the request in the spring of 1918, the rifles, for unknown reasons, were not sent. The governor made a trip to Washington and met with the chief of ordnance, who explained that the rifles would be shipped as soon as Mississippi filed a new requisition and submitted a bond for the weapons with a surety company. After the governor returned to Mississippi, he took these steps, but the rifles did not arrive at the office of the adjutant general until the spring of 1919—after the Armistice.[50]

This system of using home guards to replace the departed National Guard was not the only response states made to the demands of the Great War. The term "home guard" usually referred to a locally organized militia force, sometimes having a statewide obligation, but more often subject to employment only within their own county. For arms, these forces usually relied on their own resources, or the federal government. But many states did not follow this route to replacing their National Guard. Since the term "National Guard" had supplanted the term "organized militia," many new National Guard companies attempted to form after the United States declared war, often with the intention of being taken as a unit into federal service, as had been the case with Volunteers. Organizing these forces was difficult; first, because of a lack of understanding of War Department policy toward new National Guard units and, second, because federal conscription had removed most eligible recruits. On August 24, 1917, the War Department issued a circular letter to all state adjutants general that clarified its policy toward National Guard units that had been created after the draft of the National Guard on August 5.[51]

In it, the War Department advised those states that had not created all of their allotted National Guard units under the *National Defense Act of 1916* before the draft of the National Guard, that they could continue to organize the remainder of their quota of National Guard units if they wanted. The War Department would only arm and equip the units after the needs of the Regular Army, the National Guard already in federal service, and the National Army had been filled. Once the new National Guard units were formed and armed, the War Department reserved the right to leave them in state service, use them in federal service, or draft them into the army, at its discretion.

This policy only affected the states of Minnesota, New Jersey, Iowa, Virginia, and Indiana, all of which had been organizing new National Guard units under the prewar quota when the draft of the National Guard came on August 5, 1917. The men who joined these newly formed units did so in the expectation that they would be taken into federal service as soon as organization was completed and that by that route they would see action in France alongside their friends. However, in the fall of 1917, the War Department realized that it simply did not have the resources to equip these new National Guard units, which were far behind in their training anyway, and so issued a new decision on November 26, 1917, which, for all practical purposes, put these new units out of their misery. The new policy spelled out the War Department's intention not to call these National Guard units formed after the draft into federal service, and instead to use Selective Service as its means for raising manpower for the war.[52] As a sop to those five states which had already begun creating new National Guard units under the old quota, the War Department informed the adjutants general of those states that the men in those particular units would not be liable to conscription. This was a small concession to the men in those units, who had joined expecting to go to war with other men from their community. The governors of Indiana and New Jersey maintained the right of their units to be drafted into federal service, but the decision had been made.[53]

In spite of these announcements, some states continued to form units intended for the National Guard. In response, the War Department issued "Circular Letter Number 3" on March 27, 1918, which made clear its policy toward new National Guard units. This new policy said that National Guard units organized after August 5, 1917, when inspected and recognized by the federal government, would be furnished with arms, uniforms, and equipment only after the forces in federal service were equipped. It also stated that these troops were for state service only and would not be taken into federal service for overseas duty. The policy further stated that these troops were not part of the Army of the United States and that the men in them were liable for the draft as individuals, except for members of the units from the five states listed above. Because this policy was not clear on August 5, 1917, any state that had begun creating new National Guard units could either keep them for state service or disband them and relieve the men of their obligations.[54] Based on that ruling, Minnesota disbanded its National Guard companies. In May, 1918, the five Coast Artillery batteries that Virginia had created after the draft were called into federal service to protect installations along the state's coast. Although officially branched as Coast Artillery, their federal service had them performing watchmen duties rather than any employment as artillerymen.[55]

In spite of the difficulties that the five states had with their National Guard units that had been created after the draft of the National Guard, at least thirteen other states plus Hawaii chose to create new National Guard units for state service. Most of these states had home guard or State Guard organizations in existence, but nevertheless applied to the War Department and received permission to form new militia units following the guidelines for the National Guard. These states hoped in part to use their wartime militia as a basis for a postwar National Guard. For any of these new units to receive federal recognition would be exceedingly difficult during the war. The federal guidelines for acceptance of a unit into the National Guard insisted that, as part of the standard for acceptance, a unit must have a stable membership drawn from the locality of the armory. In the midst of the Great War, few able-bodied men of eligible age were available for joining new National Guard units. To make federal recognition as National Guard more difficult, members of the new National Guard units, excepting the ones from the previously mentioned five states, were still liable for conscription as individuals. Thus as a new unit recruited and trained for its federal acceptance inspection, its ranks were constantly reduced by conscription of its members.

With the future of the National Guard uncertain, some of the relatively more prosperous and heavily populated states created militia forces different from both home guard units and the National Guard. These were State Guard units. The Militia Bureau saw State Guards as differing from home guards in that State Guards depended on the resources of the state for equipping and arming them, whereas most home guards relied either on the individual members, or, more often, the federal government, to provide arms and equipment. States such as Massachusetts, Connecticut, New York, New Jersey, Pennsylvania, Ohio, Illinois, and Wisconsin, created state military forces that were not dependent on the federal government.[56] Not surprisingly, these states tended to be among the more prosperous in the nation. What is perhaps surprising, is that the early National Guard movement began in many of the same states. The National Guard of New York, Pennsylvania, Ohio, Massachusetts, and Illinois were the five largest in the nation. Even Connecticut's National Guard, the smallest from a state that created its own State Guard, was still the thirteenth largest National Guard.[57] These states seemingly had long been strong supporters of the National Guard, yet made little or no effort to begin the re-creation of a National Guard after the draft of the old National Guard in the summer of 1917 severed all its ties to the states. These states instead created and employed their own forces without federal assistance, and had little immediate need for a new National Guard to perform state missions.

Part of the reason for this apparent incongruity might stem from a rift within the National Guard Association that occurred several years earlier. New York, which had the largest and wealthiest National Guard in the nation, led several eastern states that opposed the movement to transform the National Guard into the form it eventually took. This New York-led group wanted the National Guard as a whole to be employed as a stopgap force, which, in the event of war, would only enter federal service long enough to allow the Volunteers to form and deploy. Jim Dan Hill, a major general in the National Guard and its unofficial historian, ascribed the lack of unity in the turn-of-the-century NGA to an "instinctive fear" of "riots and domestic strife" in the "industrial East" and a desire not to "leave the home State without the means for maintaining law and order" during war.[58] Martha Derthick, in *The National Guard in Politics,* speculated that the reason for this opposition came from the social backgrounds of typical National Guard officers in the various regions. According to her, the National Guard leadership in the South and Midwest drew heavily on men for whom their status as National Guard officers was a source of both prestige and needed income, whereas the National Guard of the eastern group drew its leaders from lawyers and other professionals who did not desire to spend long periods away from their businesses and practices.[59] Charles Dick, president of the NGA from 1902 to 1909 and sponsor of the *Militia Act of 1903*, had used his forceful personality to unite the association behind the view that the National Guard existed to fight along with the Regular Army in war, but the reasons for the uneasiness of the northeasterners had never been adequately addressed.[60]

However, this explanation does not fully explain why some states created State Guards rather than home guards or National Guards in the Great War. Although the eastern group was well represented among the states that created State Guards, so too were some midwestern states. The most plausible explanation for the tendency of some states to create State Guards that were not dependent on the federal government for arms and equipment is economics.

The states with the largest prewar National Guard that did not create a State Guard during the war were Missouri and Texas. With almost twelve thousand men each in their National Guard, they ranked sixth and seventh in early 1917.[61] Texas created six National Guard regiments of cavalry and another three of infantry for state service for the duration of the war. In addition, it authorized a system of home guards based on the authority of the county sheriff—the system in Alabama. However, neither the replacement National Guard units nor the home guard units were recorded as providing any significant service, and they were disbanded after the end of the war.[62] The situation in Missouri was a bit more complex. The provisions of the Missouri *Militia Act of 1909* were sufficient for the creation of a home guard, and no legal changes were necessary when the National Guard left in 1917.[63] The law provided that when the National Guard left the state, the governor could create a replacement militia, to be known as the Missouri Home Guard, which was to be disbanded immediately upon the return of the National Guard.[64] The departure of the National Guard for the Mexican Border had led to no calls for a new militia, but when the United States entered the Great War, Governor Frederick D. Gardner feared violent demonstrations against the war would arise among the state's large German population.[65] Each county set up a Council of Defense soon after the declaration of war, and some councils addressed the threat of antiwar agitation on their own. Kansas City businessmen began drilling and soon created military formations under their council's approval. The council in Cass County, just south of Kansas City, created a band of marshals to oppose any pro-

German elements in the county. This tended to frighten citizens who feared this local force with a vague mission.[66] The catalyst for the state government to take a more active role in local defense came in July, 1917, just before the entire Missouri National Guard of six regiments was drafted into the army. A strike in the mining region around the town of Flat River, located about fifty-five miles southwest of St. Louis, in St. Francois County, turned violent. The unrest took on antiwar overtones, which were later blamed on the "infiltration of foreign workers."[67] In response to that violent strike, one battery and one troop from the 4th Regiment, Missouri National Guard, were sent to restore order in the area, where they remained for two weeks.[68] The incident demonstrated to state and community leaders that a potential for disorder existed. But with the draft of the National Guard the following month, the state needed to create its new force.

After the "riot" in St. Francois County, the county chairman wrote to the State Military Department for instructions on forming a home guard. A resident of neighboring Madison County wrote to his council in August, 1917, urging the formation of a home guard lest similar violence occur there. The chairman of Adair County, George Bailey, in the northern part of the state, feared that idle men would be susceptible to antiwar violence, but antiwar sentiment was so prevalent in his county that "it is mighty hard to get local officials to do their duty along this line" of suppressing antiwar activities. Lack of a clear law authorizing home guards led many to believe such forces were not legal, but county leaders decided to form the force anyway. The mission of these county forces was usually described as preventing "the sinister possibilities of riot, rebellion, sabotage."[69]

Responding to the concerns of citizens, as well as to the reality of home guard units forming without state direction, the state government directed the Adjutant General to form the Home Guard from units already in existence, as well as units that might seek recognition in the future. At its peak, the Missouri Home Guard consisted of five regiments, six separate battalions, and sixteen separate companies. All of these were infantry. In addition, the force contained one separate cavalry troop. On paper, the Home Guard contained six thousand men, mostly in the infantry. The upper age limit for joining was raised to fifty, "and the rigid physical qualifications required for enlistment in the National Guard were not insisted upon."[70]

Attendance at drill averaged between 80 and 90 percent throughout the war. In the initial enthusiasm of the spring and summer of 1917, when rumors of German agents and sympathizers abounded, units held as many as three drills per week. They suffered a shortage of instructors due to federal inductions, but soon retired National Guardsmen who were too old or physically unfit for active service with the Army of the United States took over the task of instructing the Home Guard. In addition to former National Guardsmen, veterans of the Spanish-American War, graduates of military schools, and even a few veterans of the Civil War, assisted in training the new militia.[71] Although able to solve the instructor shortage, the state was forbidden by law to spend appropriations originally passed for its National Guard on the Home Guard. Over half of the last biennial appropriation for the National Guard was returned to the state treasury, a loss of $265,000 that the new militia needed. In its place, citizens subscribed over $300,000 for the force.[72] Despite the donations of money, most men in the force provided their own uniforms, while most weapons and equipment were furnished by local communities. To augment the weapons members could borrow, the state managed to get 1,200 rifles from the federal government.[73]

In August and September, 1918, the First and Third Home Guard Regiments from St. Louis held five-day camps of instruction in the suburbs using borrowed tents. The Adjutant General believed they did a great service "in keeping alive the spirit of patriotism and in stamping out any semblance of disloyalty on the part of certain elements in our population."[74] Simply by existing and showing themselves, the Home Guard regiments performed their main function as far as the adjutant general was concerned. The Home Guard by its presence kept potential opponents of the war from expressing their sentiments too openly. Still, despite such an ominous role for the Home Guard, it appears to have done little but march and train, which may have been enough to keep local opponents of national policy quiet.[75]

In addition to intimidating potential dissidents, the Home Guard provided a more concrete service to young men of Missouri. The legislature required that the Home Guard make pre-induction training available to men subject to the draft. Unlike most states, Missouri kept statistics on those men after they entered federal service. State records showed that about 90 percent of the drafted men who had received such training became noncommissioned officers upon reaching federal training camps.[76]

For those who took advantage of state military training, the benefits were substantial.

The Home Guard regiment from Kansas City, the Second Missouri, saw a week of active service when a strike against the Laundry Owners' Association by laundry workers in that city grew into a general strike of six days. More than twenty-five thousand workers took part in the strike.[77] The mayor of Kansas City was having his lunch in a downtown restaurant when rioters attacked it.[78] The arrival of the Home Guard ended the violence, although tensions in the city center remained high for several days. Streetcars could operate only under the protection of the Guardsmen.[79] In April, after the strike ended, the Second Regiment was converted into a National Guard regiment in recognition of its effective service in Kansas City. The Second Regiment, Missouri Home Guard thus became the Seventh Regiment, Missouri National Guard. Because it was not to be drafted into the Army due to War Department policy, it remained available to the state for service.[80] Although the state created and maintained a Home Guard of some six thousand men, only the Seventh Regiment performed service during periods of unrest. As the Seventh, the regiment returned to Kansas City during a strike of streetcar workers in December, 1918.[81] The Seventh also saw service in the fall of 1919, when the mine workers went on strike. The state took over the operation of the mines and used the National Guard to protect the replacement workers. The National Guard remained on duty for two weeks.[82]

Most likely the reason for the lack of official employment of the Home Guard is that it probably disbanded shortly following the Armistice. Records for that summer of 1919 show only the Seventh Regiment holding their weeklong "Camp of Instruction" at the state rifle range near the town of Nevada.[83] During the war, if the Home Guard did little more than demonstrate, it showed the people of the state that the Council of Defense in each county was ready to deal with antiwar sentiment.[84]

At least one author claimed that the state had a lack of enthusiasm for the war by January, 1918, and that widespread unrest among labor that winter had brought war production in the state to a halt.[85] At the time of the General Strike in Kansas City, thirty thousand machinists struck on the other side of the state in St. Louis. In the first quarter of 1918, only New York City had a greater number of strikes.[86] State officials feared a link between labor unrest and opposition to the war. The state made plans to maintain lists of all draft-age men and their attitudes toward the war.

Nothing came of that idea, probably because the Home Guard was a more successful plan.[87] It gave citizens who supported the war, many from the German-American population of Missouri, a means to publicly show their support, and at the same time to intimidate lawfully those people who did not support the war.

The lack of employment of the Missouri Home Guard suggests a lack of trust on the part of public officials for any organized militia that did not bear the federal stamp of approval that the name "National Guard" implied. The prosperous states of southern New England illustrated the possibilities of wholly state military forces in a region with a geographically compact population, great economic and social resources, and a long militia tradition. Although military ardor is usually associated with southern states, several northern states had strong militia traditions that rivaled the South's. Perhaps not as large a percentage of the population desired to bear arms in the local militia, but the higher population density and wealth of the North allowed militia units to thrive in some areas. Massachusetts, to which the modern National Guard traces the origins of the American militia tradition, also had some of the strongest militia traditions in the early twentieth century.

The Massachusetts Home Guard—later called the State Guard—proved better organized and equipped than most. The formation of the Massachusetts Home Guard began more than two weeks before the United States declared war against Germany. On March 22, 1917, Governor Samuel McCall sent a special message to the General Court—state legislature—asking for legislation to create such a force. The governor acknowledged that the *National Defense Act of 1916* forbade states from maintaining troops other than the National Guard during peacetime but explained that the clause allowing state police or constabulary gave ample authority for a home guard.[88] In the special message, Governor McCall explained his concern for the "bridges, water powers, [and] factories" that would be likely targets in the event of war.[89] He recommended that the age and conditions of enlistment in the Home Guard be set so as not to interfere with recruitment of the National Guard, which was then filling to war strength in anticipation of participation in the war in Europe.[90]

The General Court acted quickly on the governor's proposal and on March 22 authorized him to appoint a board of five members to plan a home guard. All board members had formerly belonged to the old Massachusetts Volunteer Militia—the state's organized militia before the *Na-*

tional Defense Act of 1916 brought all state militia into the National Guard. The state law creating the Home Guard passed on April 5, 1917—one day before the declaration of war.[91] The "Home Guard Law" authorized a militia force of volunteers who were inhabitants of the commonwealth over the age of thirty-five years, younger men who through marriage or dependents were not liable to serve in the federal forces, or men physically unfit for federal service.[92] The law specified that Home Guardsmen were to have the power of police except in regards to civil process; they could not serve warrants. The court authorized $200,000 to be spent on the force.[93] Although Home Guardsmen were to receive no pay for weekly training, they were to receive pay at the same rate as National Guardsmen when called out.

Among the companies of the Massachusetts Home Guard was the Richardson Light Guard of Wakefield. As with many Home Guard companies, the Richardson Light Guard had been formed to replace the National Guard company of the same name that had been mobilized in spring 1917. Wakefield, a town of about fifteen thousand people located ten miles north of Boston, contained an industrial base of wicker furniture and shoe manufacturing, with farming in the outskirts. The Richardson Light Guard traced its roots to 1851, when the town was still called South Reading. The town's militia company had fought in the Civil War and, by regrouping as Company A, 6th Regiment Massachusetts Volunteers, the Spanish-American Wars. After the passage of the *National Defense Act of 1916*, the company became part of the National Guard, dropping its old designation as Massachusetts Volunteer Militia and losing three men in the process.[94]

When Wakefield's National Guard company entered federal service in 1917, the town formed a new company of seventy-five men, most of whom were exempt from federal service because of age or physical disability. Two months later, when the company petitioned to become part of the new Home Guard, fifteen men opted to leave owing to a lack of enthusiasm for duty outside of the town.[95] Although the Massachusetts Home Guard was not liable for service outside the commonwealth, all units were liable for service anywhere within it. Wakefield's company became Company H, 12th Regiment, Massachusetts Home Guard.

Through the war, and for many months after, the company drilled in the town's armory at least one evening per week. On a few occasions, the men spent a weekend bivouacking and performing company drill at a farm in north Wilmington. In addition to training themselves as a company, the Wakefield men, following directions from the state adjutant general's office, began a program of preinduction training for local selectees. Along with the entire Massachusetts State Guard, the Richardson Light Guard of Wakefield spent July 25–29, 1918, at a training camp in Framingham. The company and its commander won high praise from then-lieutenant governor Calvin Coolidge when he inspected the State Guardsmen. That fall, on the day following the Armistice, the company participated in the Boston parade with the remainder of the regiment. The Richardson Light Guard returned to Boston, along with the entire State Guard, the following April to welcome home the Yankee Division from Europe.

Another company came from Concord. Perhaps no other town in America is so identified with the militia tradition. In 1915, two years before the United States entered the Great War, a local doctor formed a self-styled "Minute Man" company. Most members were over military age, but they still joined, expecting to be used "as a feeder for the regular service."[96] Most members came from Concord, but some traveled from surrounding towns to join the Minute Men. The company drilled without state recognition or support from 1915 into 1917. The Minute Men also served as a vehicle for expressing ideas related to national defense in the years before the United States entered the Great War. In early November, 1916, the Concord Minute Men at their regular Friday night drill assembly, issued a proclamation for the adoption of compulsory military training. The group then sent copies of its resolution to all governors, congressmen, cabinet officials, and President Wilson.[97] That same month, the Minute Men held a series of public lectures at the armory. Subjects included the experiences of the Massachusetts National Guard on the Mexican Border, the benefits of universal military training, and other military subjects.[98] Clearly the Minute Men were more than just a drill society. When the local National Guard company entered federal service in 1917, the Minute Men of Concord applied for, and received, a charter in the new Massachusetts Home Guard.

As with the Richardson Light Guard of Wakefield, the Concord Minute Men lost several members when they became part of the Home Guard with an obligation to serve anywhere in the commonwealth.[99] As a Home Guard company, the men of Concord were able to occupy the local armory and to receive weapons from the state. They met for two hours each week for an evening of drill. In the summer of 1918, the Concord company joined the entire State Guard at camp in Framingham. For the members of the company who stayed to the end, this period would be remembered as the apex of the company's readiness. Most members had been in the company since it began. None were "boys," and they enjoyed the manly pursuits of military life.[100]

In Holyoke, a small city on the Connecticut River, the mayor called for volunteers for what was termed a "Civic Guard" for home defense after the local National Guard company was inducted in the summer of 1917. Some six hundred men initially volunteered, with enough accepted to form two companies. The men of the Civic Guard drilled every Monday night at the state armory in town. They furnished details for all funeral services for soldiers and sailors returned to Holyoke for internment. On one occasion, the companies raised money to send to the town's National Guard company, then "D" company of the 104th Regiment, in France.[101]

Because the town's Civic Guard had no standing in state law, Holyoke organized a Home Guard company in October, 1917, with thirty-three of the men in the town's Civic Guard companies joining the new Home Guard company. With this core, it was quickly recruited to its full quota of sixty-five men and three officers. Records were kept on the prior military experience of the members of this company, and the information provides an example of the military experience civic leaders could draw on in the Great War. The men were of several nationalities, most were beyond draft age, and over half had seen service in the army, National Guard, or in the former Massachusetts Volunteer Militia. Four had participated in the Punitive Expedition into Mexico, and another three had seen service in the Spanish-American War. But American military experience did not exhaust the martial background of the Home Guard company of Holyoke, for in its ranks could be found a veteran of the Boer War, another who had served in the French army from 1914 to 1917, a former French Foreign Legionnaire, two who had served in the German army before the war, another two from the British army, and one man who had served in the Canadian Volunteers.[102]

In February, 1918, the Holyoke company underwent its State Inspection Drill by the State Inspector. He remarked on the positive aspects of the company. He said that the officers were efficient and experienced, noting that the men were "well set and soldierly in appearance." However, he also listed their shortcomings as soldiers of the commonwealth.

He noted that the sergeants had no belts or side arms, only a few of the men had regulation shoes, and the company as a whole made too much noise coming to order. Still, he closed by noting that the men performed their manual of arms well and that, in balance, the company demonstrated its competency at most of the rated tasks.[103]

An event outside of Massachusetts, indeed beyond the borders of the United States, drew elements of the Massachusetts Home Guard across an international border to assist a stricken neighbor. The American militia had refused to cross the Canadian border during the War of 1812 but would cross on a humanitarian mission in 1917 without hesitation. On December 6, 1917, at 9:05 A.M., two ships loaded with high explosives, the *Mont Blanc* and the *Imo,* exploded in the harbor of Halifax, Nova Scotia. Word reached Governor McCall about an hour and a half after the explosion. With details and the extent of the damage unknown, the governor immediately wired the mayor of Halifax asking for information and promising him that "Massachusetts stands ready to go to the limit in rendering every assistance you may be in need of."[104]

Early that afternoon, a hastily convened meeting of the Committee of Public Safety, a wartime committee responsible for order, with sixty of the one hundred members present, decided to begin a relief expedition. The committee turned to the Home Guard to provide the commonwealth's assistance to Halifax. The chief of the Medical Department in the State Guard, Dr. William A. Brooks, who also held the position of acting surgeon general for Massachusetts, quickly began organizing a hospital unit to be sent north. The receiver of the Boston and Maine Railroad promised to provide a special train for the relief forces within thirty minutes after being notified that the force was ready to go.[105]

The train left Boston's North Station at 10:00 P.M. on December 6. Because of a snowstorm throughout northern New England and the Maritimes, the train did not reach Halifax until the morning of December 8. When they arrived, they found the streets of the city full of debris and snow. They also discovered that they were the first relief force to arrive in the stricken city. The hospital unit set up its wards in the partly wrecked Bellevue Building, receiving some assistance from a group of Canadian soldiers who had been in the city when the disaster struck. Also lending assistance was a detail from the *Old Colony,* an American freighter that had been on its way to Britain for delivery to the British government but had instead received damage from the explosion. By evening, the Home Guardsmen had a working hospital consisting of an operating room and a ward of one hundred beds, with sixty patients. They raised the U.S. flag over the hospital the next day, but they were wholly a state relief force.

Although the medical detachment from the Home Guard provided the most immediate relief, it did not comprise the entire effort from New England. Massachusetts sent a relief ship on December 9, which carried, among other things, ten vehicles and drivers. The Massachusetts Red Cross also established a hospital in the stricken city. Volunteer doctors and nurses from Maine arrived a few days after the Massachusetts Home Guard. In the final tally, the explosion left about 1,800 people dead and another 10,000 injured, of whom many later died. With the cold and snow arriving even before the official start of winter, more than 2,500 homes, as well as many other public and private buildings in the city, were destroyed. In 1919 dollars, the total estimated cost came to more than thirty million dollars.[106]

Until the fall of 1919, the Massachusetts Home Guard performed few other missions besides training. Two companies spent most of December, 1917, at areas designated by President Wilson as "danger zones" in the commonwealth, mostly near docks.[107] They were relieved by U.S. Guards on Christmas Eve.[108] That spring, the Home Guard officially became known as the State Guard, which better reflected its commonwealth-wide obligation.[109] Later that spring, the company from Easthampton spent two weeks on active duty responding to unrest in that town. Fall of 1918 saw the State Guard involved in the struggle against the influenza epidemic. Most of its duties involved establishing field hospitals and transporting doctors and nurses. After the Armistice, the State Guard shrank but continued as Massachusetts's main organized militia. The company in Franklin, near the Rhode Island border, responded in February, 1919, to an explosion in the town, and was called out in response to civil unrest that summer. The company in Orleans, near the tip of Cape Cod, searched for a missing person in spring 1919.

But the State Guard was not the only organized militia in Massachusetts. All three of the southern New England states had militia organizations, some of long standing, which despite the provisions of "Section 61" of the *National Defense Act of 1916*, had remained outside of the National Guard. In Massachusetts, the Ancient and Honorable Artillery Company traced its roots to shortly after the settlement of Boston in 1630.

The company had long been a social rather than a military organization, but it still kept the trappings of and state status as a militia unit. The company had hundreds of members, who annually elected the company's officers. They counted Great Britain's King George V as an honorary member. Such a group could not remain aloof with the nation at war.

A provisional company, a self-selected group of about one hundred Ancients who desired more military-type training, had been drilling each week since early 1916. This group later expanded to a battalion. Contributions were solicited from all Ancients to equip the battalion. When the State Guard was organized, many of the Ancients who had been drilling temporarily left the company to become officers in new organizations.[110]

The Ancient and Honorables, with a long and proud military past, sought a more direct role in the war. With the entry of the United States into the Great War, the commander of the Ancient and Honorables, Captain Lombard, offered to the adjutant general of Massachusetts the services of the company to organize and equip a field artillery battery for active service. The personnel for this battery were to come from the Ancients and others whom they selected. Captain Lombard had not realized that the Great War was not to be fought as had previous American wars and that the type of service he offered to provide had become obsolete.[111]

The adjutant general declined his offer, but for many months the state did borrow the company's machine guns, which it had bought years earlier with its own funds. Unable to raise a company for active service with the army in Europe, the Ancients set about raising money from within their membership to equip a State Defense Battalion, to be manned by company members.[112] To arm this group, the Ancients bought one hundred Krag rifles, along with other gear, from the War Department. Although the Ancients paid for the equipment, the War Department delayed shipping the arms and equipment. The Ancients wanted the items for a parade that had been scheduled for that June, and used its friends in Congress to put pressure on the War Department to ensure that the equipment arrived in Boston on time. The Ancients wielded considerable influence; they were able to bring political pressure to bear on the War Department to issue equipment to the company for a parade, while the War Department was trying to raise and train an army to fight a world war.

Despite their inclination to present a martial turnout for parades at the expense of the national war effort, the Ancients took their standing as an organized militia of the commonwealth seriously. Members were reminded by the company's leadership that they were to wear only the olive drab uniform, which they bought themselves, during the war, in accordance with the orders of the president. They also needed reminding that, as members of the military forces of the commonwealth, they were not free to enlist in any other home guard, State Guard, or such other state militia units. They could, however, instruct or help form such units.[113] Still, the Ancient and Honorable Artillery Company, although far larger than any National Guard or Regular Army company, was still only a single militia unit and constituted only a very small part of the Massachusetts State Guard. But in another southern New England state, Rhode Island, other independent militia organizations would have the greatest impact on any state militia forces in the Great War.

Rhode Island has always had an organized militia separate from the National Guard in the form of its Independent Chartered Military Commands, plus a few that were unchartered. These were militia units, some ancient, some not so ancient, that existed within the state. Like many militia groups before the rise of the National Guard, these were more social than military organizations. With the nation at war, they were encouraged, particularly the groups in Providence, to increase their numbers and acquire more supplies. The First Light Infantry Regiment received most of its financial aid from past members and soon recruited to a strength of five hundred officers and men in five companies. The state offset some of the cost of their uniforms but supplied rifles for only four hundred men. Another unit, the Newport Artillery, increased to one hundred men and provided its own equipment with money raised through private subscription. Other nominal artillery units such as the United Train of Artillery, the Bristol Artillery, and the Kentish Guards also recruited to more than a hundred men each, but these groups received no equipment. Soon all units began a program of drill and other military instruction.[114] The adjutant general recalled a past commander of the Newport Artillery, Col. Alvin A. Barker, retired, to command a State Guard to be created from the military organizations already in existence. Colonel Barker received no pay for his work.[115]

The adjutant general said the expansion of the Chartered Military Commands stemmed from an increased desire in the state for an adequate state military force during the war. He suggested that the state create a regiment of infantry using the Chartered Commands as a nucleus for at least a battalion in Providence, with other companies at Newport, Bristol,

Warren, and East Greenwich. He also hoped to see companies in Woonsocket, Pawtucket, and Westerly, filling all of the state's armories. His office drafted the plan for the General Assembly to enact into law.[116] Although many of the Chartered Commands had at least one hundred men, the plan as suggested by the adjutant general outlined companies with a strength of eighty men.[117] This meant extensive restructuring for most of the Chartered Commands, which more often than not designated themselves as "companies" but were in reality more battalions or even regiments. The Newport Artillery "company," to use one such Command as an example, had a colonel as its commander, who was assisted by a lieutenant colonel, major, and few captains, plus a surgeon with the rank of captain. Others commands, whatever their name, tended to have a similar organization.[118]

The legislative orders for creating a State Guard came on May 9, 1918. In reality, this law recognized the Chartered Commands, as well as a few unchartered militia groups, as the State Guard of Rhode Island. Nevertheless, as the State Guard, all militia organizations had to conform to a basic plan and follow a few rules. Members had to be over eighteen years old,[119] enlistments were for the duration of the war plus six months, and the total number of companies the state would accept was limited to thirty-six, but not less than eighteen.[120]

"Section 4" of the General Order said that the adjutant general could organize the Chartered Organizations as State Guard units but that none of their rights and privileges could be affected.[121] In other words, their positions as Chartered Organizations came before their new status as State Guard. In effect, these organizations maintained two separate standings under state law, their Chartered Organization standing temporarily superseded by their State Guard standing. The Act also set aside fifty thousand dollars for the use of the force.[122]

Colonel Barker began changing the Independent Chartered Military Organizations "and all recognized constabulary commands" from elite social clubs into real militia suitable for service as a "Home Guard force."[123] The "recognized constabulary commands" he referred to were units similar to the Chartered Organizations but had never been chartered by the state. The state had "four Constabulary Commands which had previously [been] reported to the War Department as authorized to bear arms, [and] were tentatively recognized."[124] For the State Guard, the state recognized any company with one hundred enlisted men, and then inspected them for mustering into the State Guard.[125] Some of the men had earlier signed up for three-year enlistments. Those who had done so kept the same terms, but all new recruits were for the war plus six months.[126]

The state bought much of the equipment for the force, including Colt .38 revolvers from private dealers. Seemingly just as important were "trumpets and cords, and hat cords" for the force. If Rhode Island was to turn out a militia, it had better at least look like a military force, complete with violet cords on their campaign hats. At great difficulty, given that the war effort gobbled up most martial items the economy could produce, the state secured sky-blue overcoats and capes of the "old pattern," even though most stocks in the country were exhausted.[127]

Units solicited funds for uniforms in their home communities. The federal government supplied five hundred .45 caliber Springfields, and another 250 Krag Jorgensen caliber .30s. The Newport Artillery had one hundred U.S. Magazine Rifles, caliber .30. In addition to what the federal government could supply, the state owned another six hundred Springfield .45s. As this did not satisfy the state's hunger for weapons for its force, they also ordered 1,400 of the "Russian" rifles. The issue of these rifles necessitated the return of the Springfields and Krags. Eventually each company received eighty "Russian" rifles and twenty Springfield .45s. The Machine Gun Detachment was armed with fifty of the "Russian" rifles, in addition to their machine guns.[128]

The training of Rhode Island's new State Guard force reflected the independent nature of its origins. Although these units did train, their methods of training differed from what home guards of most other states experienced. Indeed, the training of the Rhode Island State Guard differed from anything either the National Guard or Regular Army had done in years. The State Guard did follow the National Guard's marksmanship program, but no qualifications were required and no medals or trophies awarded. Companies marched and camped on their own resources. One company marched to Fort Kearney, in the state, in cooperation with the regular garrison stationed there. Another did tactical training with the cadets in ROTC at Brown University in May, 1918.[129] That fall, in order to impress upon (or simply to impress) the resident of the state capital that the state had a well-organized militia during the war, the State Guard marched through Providence in heavy marching order, with the governor there to receive and review them. Ready for trouble or not, the State Guard at least looked ready.

On Thanksgiving Day, shortly after the parade and less than two weeks after the Armistice, the State Guard staged a combination training exercise, publicity stunt, and fund raiser. This event consisted of a sham battle reenacting the Canadians taking Vimy Ridge in France. Before the twentieth century, the reenactment or sham battle open to the public had been a common event, especially with militia units, but also in various training corps and even in the Regular Army. But since the radical changes that had swept through the American military establishment between the Spanish-American War and the Great War, the practice had been largely abandoned by all. But here at the end of the Great War, with most manpower in short supply, and the federal forces having absorbed most equipment and ammunition, the Rhode Island State Guard was able to train like a militia unit in 1880.

For this battle, the State Guardsmen and cadets dug trenches through the Narragansett Trotting Park. The ROTC cadets came from Brown University to assist them in this undertaking. Part-time soldiers from the upper crust of Rhode Island society and Ivy Leaguers from Brown wallowed in the dirt to show their martial skill for the people of Providence. Members of the State Guard played both the Canadians and the Germans. To make the event more spectacular, the State Guard managed to scrape together over two thousand dollars to purchase explosives, despite the wartime demands on munitions. The Guardsmen placed the charges in holes and ignited them electrically: dazzling effects for 1918, and a testament to the organizational and economic resources of the men of the State Guard.[130]

During the reenactment, a premature explosion injured two Guardsmen and burned a few others. The stated purpose for the sham battle was to raise money to allow the force to purchase woolen uniforms for the coming winter, but expenses were too heavy and attendance was too light. The State Guard barely broke even on the sham battle, although the training was supposed to have been quite valuable. Still, had their resources been put into purchasing the needed uniforms outright, they could have outfitted most of the force. But they would not have had the fun of taking Vimy Ridge from the Huns while never leaving Rhode Island.

A report of a training camp held at Quonset Point, August 30 to September 2, 1918, gives additional insight into the manner in which the Rhode Island State Guard approached its training. The daily schedule shows the men awakening at 5:40 A.M., drilling from 8:00 to 10:00, then having a swim party every day. NCOs attended guard mount school in the afternoon. Throw in a bit of sentry duty, KP, and some evening entertainment, and the day was complete. Not a bad time at all for the men attending, but was it the most efficient use of time?[131]

Fortunately for Rhode Island, as well as the men in the State Guard, the state never had a violent incident needing the State Guard. Rhode Island was a small and prosperous state and weathered the war years well. The only use of the force came when the Third Battalion's sanitary detachment did duty for the influenza epidemic from October 12 to 31, 1918, at Pawtucket.

For the nation as a whole, the year and a half between the declaration of war and the Armistice was relatively calm. Having an organized militia usually provided a measure of comfort to the states that created them, but the homefront experienced few incidents during the war requiring a response from state military forces. Eventually, some twenty-seven states created a replacement militia for their National Guard.[132] Altogether, the War Department estimated that 79,000 state troops were raised during the Great War.[133] Other states created no forces and instead depended on federal troops in case of riot, strikes, or natural disaster. For those areas not protected by a viable home guard, the War Department organized a force of twenty-five thousand men in forty-eight battalions called the United States Guards. This force was formed from men who had been drafted and trained but for physical reasons were unqualified for overseas service. The U.S. Guards relieved deployable men from guard duty at some 338 industrial and strategic points around the country.[134] The end of the war proved to be the beginning of a period of social dislocation. With the future of the National Guard in question, states with a well-maintained militia found themselves at a distinct advantage over states without a reliable force.

CHAPTER 3

Postwar Adjustments

On November 14 and 15, 1918, the National Guard Association (NGA) met in Richmond, Virginia, for its first meeting since March, 1917.[1] The previous meeting had been held at a time when the National Guard had just returned from the Mexican border and was about to be recalled into federal service for the Great War. The passage of the *National Defense Act of 1916* had fulfilled most of the goals for which the National Guard Association had fought for decades, but when the National Guard entered federal service, the association temporarily ceased to function. With practically all National Guardsmen on active service, members had a war to fight and had no time for running the association. In addition, legally they served as soldiers in the Army of the United States, losing all standing in the National Guard in the spring of 1918, when the army abolished all distinctions between the Regular Army, National Guard, and National Army. However, by the fall of 1918, some adjutants general grew increasingly nervous over the future of the National Guard. In order to ensure the re-creation of the National Guard along similar lines as before the war, the National Guard Association needed to begin functioning as a coherent pressure group before the federal government began the debate on the future of the National Guard.

Brig. Gen. Harvey J. Moss, the adjutant general of Washington State, and president of the National Guard Association, began the drive to reorganize the NGA in the late summer and early fall of 1918. By chance, the dates chosen for the first conference fell three days after Germany signed the Armistice. The conference drew representatives from only thirteen states, but the excitement of the time led to a slightly giddy atmosphere at the conference, so much so that the delegates may have missed the message in the speeches they heard.

The most important purpose of the meeting was to elect temporary officers for the association to allow it to function again. The main speaker was Brig. Gen. John S. Heavey, chief of the Militia Bureau. Heavey was not a National Guardsman but a Regular. To the National Guardsmen assembled in Richmond he made no secret of where he placed his sympathies. He told them not to claim too much credit for the recent victory in Europe. True, he said, the National Guard had committed all of itself to the war and had helped the effort greatly, but its contribution needed to be put into perspective. In the greatly expanded wartime army, the National Guard amounted to only about 10 percent of the total Army of the United States. Heavey instead suggested that in the postwar military establishment, the army would be better served by a wholly federal reserve, although he conceded that it might be organized by state during peacetime. Rather than have, for example, the Missouri National Guard with its federal and state missions, a new "Missouri United States Reserves" might be the institution through which men of Missouri would train part-time in peacetime for the next war. Such a force would have no state mission, and the state would have no control over it, even during peacetime. Despite this being anathema to the National Guard Association and everything it had long fought for, the comments passed without remark from the assembled audience.[2]

Heavey talked at length about the U.S. Guards. Although a part of the army and not National Guard, the U.S. Guards fell under the control of the Militia Bureau mostly out of convenience. Like the National Guard Association, the Militia Bureau became something of a headquarters without an army after the draft of the National Guard. Responsibility for the U.S. Guards fell to the Militia Bureau by default, as the Bureau had little else to do during the war. Heavey claimed that his office organized thirty thousand men in the U.S. Guards, but their creation would not have been necessary if all the states had created a home guard. He had bitter words for states that took the attitude of "you took our state troops, you do their state duty" while the War Department was trying to win a war. He believed that internal security missions were almost by definition state missions, and he resented that some states had created no organized militia to replace the National Guard.[3]

Wartime home guards created by the states contained about 130,000 men, of whom half had eventually been armed by the federal government, while another 20,000 had been armed by the states or by themselves. Heavey believed 130,000 men to be the upper limit that the states should create for state service. Since many states created no forces at all, Heavey must have felt that some states created more home guards than they needed. To him, the complete separation of state and federal military forces would prevent a repetition of the failure of some states to create a militia during war. What the chief of the Militia Bureau proposed to the National Guard Association was the end of the National Guard. Rather than re-create it as it was before the war, he wanted a new system entirely. He saw a smaller, permanent home guard force supported completely by the states for state service and without a federal mission, and a separate federal reserve force. This new federal reserve would be completely free of state missions and interference.[4] The strength in such a system would be that in any future war the army could induct its reserve force without disrupting the organized militia in place in the states, thus the ad-hoc nature of the home guards in the Great War would not be repeated. Of course he did not mention the impact of federal conscription on any home guard in the future, indicating that he believed only men who would be ineligible for wartime conscription would be allowed to join a home guard.

Had this same speech been given at any other National Guard Association conference, association members would have howled in outrage and done all in their power to have Congress replace the chief of the Militia Bureau. But the association was at its weakest position since the early 1890s. Representatives from only thirteen states were present, and they had no National Guard in being to represent. Besides, the United States military forces had just ended the War to End All Wars. The New World had righted the wrongs of the Old World, and the members were not about to quibble over technicalities. Heavey had the good sense to end his speech by praising the inborn qualities of the American citizen soldier. He told of an American soldier in France who went over the top alone to face a unit of the Prussian Guard. The American soldier broke his rifle and bayonet through his inexperience but chased the Prussians anyway with his fists. The moral was that Americans were naturally full of fight, but they needed discipline. Unspoken but implied was that he did not see the National Guard as the vehicle to instill the discipline needed to channel that courage and pluck for the next war. With victory still in the air, the association members took no notice of his ominous sentiments and instead could only applaud his testimony to the natural fighting spirit of the American citizen-soldier.

Despite the cloud over the future of the National Guard expressed by Heavey, the association representatives began the re-creation of the National Guard Association as a viable pressure group to lobby for the new National Guard, on the same lines as it was before the war. Temporary officers of the association were elected and plans formulated for the next conference, the real first postwar conference, when a strategy for the future would be decided. For many adjutants general, re-creating the National Guard Association took on a sense of urgency with the end of the war. Without the association's lobbying power, they would be in a weak position vis-à-vis the opponents of the National Guard in the Regular Army and War Department. And most of the adjutants general, led by men like General Moss from the state of Washington, desperately wanted a restoration of the National Guard as it had existed before its induction into the army. They needed their most reliable force to respond to the domestic problems that began as the war ended.

Even before the entry of the United States into the Great War, Washington State experienced a period of labor violence that, although it abated during American involvement, would not finally end until the next decade. Several strikes, labor disputes, and civil disturbances occurred during 1916. As adjutant general, Moss had prepared to mobilize the National Guard on several occasions, but only twice had it proven necessary. A clash between a sheriff's posse and "two shiploads of members of the IWW"—the International Workers of the World, or "Wobblies"—on November 5, 1916, at Everett, resulted in the death of Lieut. Charles O. Curtiss of the National Guard Reserve, as well as six workers.[5] The Wobblies had tried to land at the City Dock, in defiance of city authorities, in order to hold public speeches. When the ships carrying the IWW returned to Seattle, Everett's Sixth Division, of the state Naval Militia, as well as some National Guard companies, remained on standby until November 7, when tensions abated.[6]

The next summer, the National Guard would again be called out to deal with the IWW in Washington. In July, 1917, the sheriff of Kittitas County, on the eastern slope of the Cascades, called for the assistance of Troop "A" of the First Cavalry Squadron in connection with the arrest of a large number of IWWs at the county seat of Ellensburg.[7] A detachment was put on duty and the arrest occurred without incident.[8] But the tension in the state, and fears that the IWW would create disorder during the war convinced Governor Ernest Lister that the state would need a home guard when the National Guard left.

The adjutant general's office had drawn up plans for a home guard before the nation entered into the war. Authorization for the force came from the governor on July 11, 1917. The state brought Walter B. Beals, a retired National Guard major, on state active duty to recruit and organize the force.[9] It originally had sixteen infantry companies, organized into one large Third Provisional Regiment. The companies followed the U.S. Army's organization tables for peacetime infantry companies. The terms of enlistment copied that of the National Guard—three years active followed by three years reserve, unless a peace treaty changed the situation.[10] Originally, state law circumscribed the use of the State Guard beyond the state's borders, but later this clause was removed, most likely through mutual agreements with neighboring states.[11]

In August, 1917, the state brought a retired National Guard colonel, William E. McClure, to duty to command the Third Provisional. The state accepted the regiment as its State Guard on November 15, 1917. To augment it, a medical department was organized in December, 1917, and a machine gun company was also organized for service in Seattle.[12] In 1918, fifteen members of the State Guard were chosen to patrol forests under the direction of the state fire warden and the Forestry Fire Association, whose normal forces had been depleted by the draft. They were picked from the hundreds of volunteers who indicated that they had firefighting experience and their own car.[13] They remained on duty in the state's forests for four weeks.[14]

Following the issuance of "Circular No. 3" by the War Department, the Adjutant General began the reorganization of the Third Provisional Regiment into a regiment of the Washington National Guard. The transformation of this unit, originally a home guard regiment, began in May, 1918, with federal acceptance scheduled for July 15. However, shortcomings in manpower and training delayed federal acceptance until September 30, 1918. After that date, the Third was a National Guard regiment, with drill pay provided by the federal government, but it remained in state control and thus stood ready to serve the state.[15] This regiment was intended to form the nucleus of the postwar National Guard.[16]

Throughout the life of this regiment, both as home guard and as National Guard, men were drafted out of it for federal service. War Department policy provided that membership in a National Guard unit organized after the draft of the National Guard did not make a man ineligible for individual draft. About 1,200 men were drafted from the regiment after recruiting began in August, 1917. Despite its new status as National Guard, federal arms and equipment were not forthcoming, though repeated requests were sent to the Militia Bureau. The federal government simply had too few weapons and not enough equipment to spare for these new National Guard organizations.

As a hedge against a change of War Department policy, and the possible, although unlikely, draft of the Third, the adjutant general maintained four companies of infantry in the State Guard.[17] Despite the presence of this backup force, Washington State's adjutant general became the main driver behind reorganizing the National Guard Association, even before the war ended. Moss had made the general call to the military department of each state to announce the meeting of the NGA in Richmond, Virginia, in November, 1918, to lay the groundwork for a postwar NGA that would push for the full re-creation of the National Guard as soon as possible.[18] As Moss had one of the larger and better organized State Guards in the nation, his desire for a fully re-created National Guard betrayed his lack of confidence in the State Guard.

The adjutant general now had under his command four State Guard companies, as well as the Third Regiment, which was officially part of the National Guard. But evidence of a lack of faith in the militia came during the February, 1919, Seattle General Strike. Seattle's mayor, Ole Hanson used the threat of federal soldiers to break the strike.[19] The state's troops in Seattle were placed on alert and held in readiness at their armories, but federal troops actually intervened in the strike. However, the unrest in Everett on October 8, 1919, and in Spokane from November 14 through 16, brought the state troops in those cities to active duty to maintain order. The near riot in Centralia, Washington, on the first anniversary of the Armistice, brought in the National Guard company from Tacoma to restore order when an American Legion parade degenerated into an attack on the local Wobblie hall.[20] The unrest in Washington alarmed many adjutants general throughout the nation, and Washington's inability to prevent outbreaks would provide impetus and a sense of urgency for re-creating the National Guard in the postwar period. The four remaining companies of the Washington State Guard ceased to exist by 1920 at the latest.

Oregon, Washington's neighbor to the south, had similar difficulties with organized labor, yet from a more chaotic beginning it created relatively efficient state forces to deal with labor unrest. With the start of American involvement in the war, petitions from many communities began flooding the adjutant general's office requesting permission to create home guard forces.[21] The state's official position on the matter was that communities could do whatever they wanted but the state itself would not be taking an active role in creating a statewide force.

The state hesitated organizing a militia because of fiscal restraints, but also because it feared that doing so would incur future claims of liability against the state. However, when the state's attorney general ruled in November, 1917, that the governor could use the state funds that had been earlier appropriated for use by the National Guard to fund a home guard, the state acted. At Portland on November 19, 1917, a battalion of State Militia, largely composed of Spanish-American War veterans who had organized in July, 1917, was extended state recognition.[22] This battalion, in the state's largest city, served for a few months as the state's only organized militia. Apparently this urban militia in the state's only port of any size, existed as a check on radical IWW elements.

Other parts of Oregon also provided opportunities for local men to band together for what they believed was a matter of protecting their communities and their nation. Local "vigilante organizations" sprang up throughout the state from grassroots enthusiasm and without central direction. County sheriffs reacted favorably to these groups and soon began to deputize them for service within their counties. Eventually, most counties in the state had some sort of vigilante organization. These County Defense Organizations, as the vigilante groups were soon renamed, had no state military status.[23] They nevertheless were called to maintain law in Portland and Coos Bay district, on the southern coast, as well as other places with labor strife.

To arm the force, the federal government in August, 1917, loaned the state some two thousand 1884 model Springfield rifles, and 50,000 rounds of ammunition. Of these, 625 were issued to the sheriff of Multnomah County, where Portland was located. The state requested even more rifles in order to issue them to the County Defense Organizations but were informed by War Department authorities that the state had already received more than its quota and would not be getting any more weapons from the federal government.[24]

The fear of enemy aliens and the IWW, plus a desire to channel the enthusiasm of the volunteers, led the governor to transform the scattered units, plus the State Militia battalion in Portland, into a State Guard that would be available for statewide service. He said he needed a State Guard because the IWW posed a threat to the smooth functioning of agriculture and industry in the state. The State Militia battalion in Portland would not be enough to deal with what he believed to be a large and well-organized conspiracy. County Defense Organizations that wanted this new obligation for service statewide were mustered into the state militia. The new force was christened the Oregon Guard, which consisted of the Portland Battalion of the State Militia, expanded to form the First Regiment with another battalion in Salem, and the former County Defense Organizations.[25] The state was careful to ensure that this new State Guard caused no new drain on the state treasury. The men who joined were enlisted for two years unless sooner discharged. Eventually, thirty-five companies were thus mustered into state service. The state bought uniforms for the First Regiment, which was based in Portland and Salem, while counties usually provided uniforms for the other units.[26]

By the beginning of 1918, the mayor of Portland, and the sheriff of Multnomah County, feared that the IWW planned to cause disruption within the city that could reach levels of violence beyond the ability of civil authorities to contain. As a preemptive measure, they placed one officer and 50 men from the Portland Battalion on duty at the harbor front beginning on January 21, 1918. This force was later increased to 100 men at the request of the Portland chief of police, who later increased it to 125 men. Privates on sentry duty received three dollars per day, while officers received standard federal rates. The men usually slept in the armory.[27] The men protected fifteen large industrial plants, dockyards, mills, and public utilities at night in six-hour shifts.[28] The militiamen watched the docks until April, when a new state military force relived them.[29]

The force, unique to Oregon, had its genesis in the spring of 1918, when the State Council of Defense came to believe that the part-time Oregon Guard would not be enough to maintain order during wartime. As a result, the Council created a full-time force of 235 men called the Oregon Military Police. In essence, this was a state police force, but organized as a small standing army and paid out of state militia funds. The Military Police remained on constant service for almost a year beginning April 1, 1918. This new, full-time force relieved the militia of Portland in performing duty at shipyards and in the grain districts in the eastern part of the state. The force also assisted in enforcement of state prohibition laws, as well as allowing the governor to fight other obstructions of the law. Throughout its existence, the Military Police had the legal status of militia on emergency service. After the end of the war, the Military police battalion was reduced to two officers and twenty-three men.[30]

In the fall of 1918, about a month before the Armistice, the state held a training course for officers of the Oregon Guard. One hundred and twenty-five men attended the four-day training session at the Portland armory. At the time, state officials believed, as did much of the leadership of the American army, that an end of the war was still far off. At the same time, the Oregon Guard underwent a reorganization that brought the separate battalions of the force into a single brigade of 37 tactical units. Part of this reorganization was the creation of four regiments, with officers electing the new regimental commanders.[31] Although the war ended shortly after these changes were made, the Oregon Guard remained in existence. After the war, it fought forest fires in the area around Portland, and some units aided civil authorities in the state, all without receiving pay.[32]

The state military also maintained an intelligence department, whose official purpose was to catch deserters and draft dodgers from the federal forces. In addition, this force, which consisted of a major and several company grade officers, but no enlisted men, had the mission of uncovering spy rings within the state. As befitting a secret organization, little record of its activities survive.[33]

The Oregon Military Police, in its much diminished form, remained on duty until February 28, 1919. The units of the Oregon Guard were mustered out in April and May, 1919. Members who met the criteria for membership in the National Guard and who elected to take the dual oath to the state and nation became the nucleus for the postwar National Guard.[34] Oregon's first postwar company of the National Guard received federal recognition on June 30, 1919.[35]

With the end of the war in Europe, the future of home guards throughout the nation had fallen into limbo, yet most states had no National Guard. Home guard forces were to exist only during wartime, but at the end of the war they constituted the only state military force in much of the nation. Disbanding these forces would leave states without an organized militia. For practical purposes, the National Guard of all states and territories had ceased to exist on August 5, 1917. State adjutants general and their headquarters staff constituted the entire National Guard during the war. Although the National Guard divisions continued to exist within the army, wartime mobilization had moved men and units between the Regular Army, National Guard, and National Army divisions. Units could be demobilized from federal service back to state service, but the individuals who comprised those units in the summer of 1917 were discharged as individuals. Upon demobilization and discharge of the former National Guardsmen from the Army of the United States, individuals were under no obligation to rejoin the National Guard. Under federal law, National Guardsmen could be held in federal service past the expiration of their enlistments in the National Guard, but on discharge from federal service, they returned directly to civilian life and not to the National Guard. But that point became moot when the attorney general ruled that when President Wilson inducted the National Guard into the army in August, 1917, individual National Guardsmen stood discharged from their state militia, and owed nothing to the National Guard upon their discharge from the army.

Some states had already begun creating a new National Guard even before the war ended, but the uncertain standing of these new units in the postwar military establishment placed these units in limbo. Most other states decided to wait for the federal government to decide on a policy towards the National Guard before proceeding to form new units. Typical of the states which had started to create a new National Guard on their own was Alabama. A few days before the Armistice, the governor of Alabama began the creation of a new National Guard strictly for service within his state. He based his decision to do so on "Circular No. 8" from the Militia Bureau, which encouraged states to begin creating new National Guard regiments but cautioned that the federal government would not be able to arm them, that members would still be eligible for conscription into federal service as individuals, and that states should form these new regiments one company at a time. The new companies were to be inspected by an army officer for acceptance as National Guard.[36] The man the governor appointed to create this new National Guard, Virgil V. Evans, informed the governor that he intended to start recruiting the first company immediately.[37] The governor of Alabama, indeed the governors of all states, had no need to fear losing this new National Guard to the National Army. General Heavey, the acting chief of the Militia Bureau, assured Senator John H. Bankhead that the War Department had no plans to send any new National Guard units overseas but considered them strictly as militia, either for state use, or for use as federalized militia within the United States if the president so directed.[38] The governor of Alabama wrote to General Heavey on October 15, telling him that his state legislature had not met since the National Guard entered federal service and thus the state had no laws for forming a new militia. He wanted to know if units of men physically unfit would be permitted.[39] The creation of National Guard units of physically unfit men was not possible, because new National Guard units still had to pass federal inspection. With federal conscription taking most able-bodied men, raising new regiments of men eligible for the higher standards of the National Guard would prove slow and difficult.

The National Guard of the immediate postwar period was a hollow force—most states had only a handful of members. The problem existed nationwide. By 1920, Congress appropriated funds for a National Guard of 106,000 men, but the total enrollment was only 56,106.[40] An example of the difficulty in raising a postwar National Guard was the situation in Mississippi, which had been allotted a force of 2,098, but could muster only 305 National Guardsmen by May 1, 1921.[41] The tension in some states between home guardsmen, who already had units in existence and hoped for them to remain so, and National Guardsmen, who wanted to rebuild the National Guard, added to the problem.

The chief proponent of reorganizing the National Guard in Mississippi was the adjutant general, Brig. Gen. Erie C. Scales. To accomplish this, he needed to recruit an almost entirely new force and did not want to compete with the Home Guard. Some Home Guardsmen foresaw an enlarged Home Guard as the organized militia for the postwar era. General Scales hoped to disband it as soon as he began recruiting a new National Guard. However, supporters of a permanent home guard soon found an ally in the U.S. Senate. On October 23, 1919, Senator Pat Harrison of Mississippi introduced a bill into the Senate that, if passed into law, would have permanently authorized the secretary of war to issue weapons and ammunition, as well as other pieces of soldier kit, to home guards or any other state military forces as each state's governor saw fit.[42] This would make home guards a permanent organized militia, for strictly local defense, rather than a wartime expedient. General Scales first learned of the proposed law in an interview with Senator Harrison published in the *Commercial Appeal* of October 23, 1919. He did not like it.

On October 27, he wrote the senator a letter in which he outlined his opposition to the bill and asked Senator Harrison to withdraw it. To General Scales, the bill would greatly frustrate his efforts to reconstitute the National Guard of Mississippi. The Home Guards of Mississippi had not been able to recruit to strength during the war, and he saw a postwar continuation of the Home Guard as potential competition for recruits for the National Guard.

For General Scales, the National Guard represented the only feasible organized militia for Mississippi and he protected that role with a vengeance. In addition, he stated, "certain citizens of the state claim to have seen sufficient service brought about on account of the war and in many instances are rather inclined to organize companies for home defense only."[43] Many former National Guardsmen had no desire to fight in another war, yet still desired to be in an organized militia. With a separate organized militia without a federal mission, these citizens might opt for Home Guard membership rather than National Guard. General Scales feared that the National Guard, which had fought for so long for an official role in the defense establishment of the nation, would be replaced by a purely state constabulary that would have no federal mission. He also expressed his fear that the equipment given to Home Guards would fall into the hands of "irresponsible persons and these arms might be used to equip strikers, mobs, and similar organizations."[44] He noted that twenty-five counties in his state were authorized by the War Department to have National Guard units and that the state would assist any other community that went through the proper procedures to get a National Guard unit. He added that "if all concerned would get the Home Guard organization off of their minds and out of the way, National Guard units could be recruited to the authorized strength and armed and equipped."[45]

In a very conciliatory reply of November 3, 1919, the senator mentioned the numerous letters he had received from Mississippi "telling [him] of the very serious situation there so far as the negroes are concerned[; that he] felt that Congress ought to do something for the people's protection and safety."[46] The senator explained that he wished every community in Mississippi had a company or a platoon of National Guard, but because of the size or age of the population in some places this would not be possible. In these communities, "not just in the South but in some other sections of the country," he hoped to allow the people to form home guards for their protection.[47] He further stated that he supported the National

Guard fully and did not see his legislation as competing or interfering with the reorganization of the National Guard. Increased federal control over the National Guard had raised the requirements for membership in the National Guard, and the senator feared that some smaller communities would not have enough young and healthy men to meet federal requirements for the establishment of a National Guard unit. This would leave these communities without an organized militia. He asked General Scales to discuss the matter with the governor and reconsider his position.[48]

The general did not delay in replying to the senator. On November 13, 1919, he confirmed that his office, along with the governor, had studied the bill proposed by the senator and he still believed that it was a bad idea. He explained that the National Guard held the responsibility in peacetime for the missions that the bill would give to home guards. Besides, he wrote, "it is the opinion that all loyal and patriotic citizens should rally to the call for the reorganization of the [National] Guard and prepare a trained force for the protection desired."[49] He also felt confident that, although his staff of five experienced officers was meeting difficulties in this task, the "recent activities of the I.W.W.'s in Washington state will be the means of speeding up the reorganization."[50] The unrest in Centralia had occurred only two days before the adjutant general drafted his reply. Facing the threat of domestic insurrection and riots, the general believed that the federal government, as well as state governments, would begin the reorganization of the National Guard in earnest for traditional militia functions.

With the absolute hostility of the man who was to administer the Mississippi Home Guard—the state adjutant general—the small force disbanded as Scales began the reorganization of the National Guard. The Mississippi Home Guard of the Great War, unwanted by the adjutant general and never a viable force, ceased to exist by 1920. Scales feared a permanent home guard would mean an end to federal support for the National Guard, with a new federal reserve formed to take its federal mission, or the National Guard re-created and taken away from the states, making a home guard maintained with state funds necessary. States like Mississippi were ill-prepared to maintain an organized militia from their own funds. Most states supported the structure of the National Guard as it existed before the Great War because it gave them a militia force paid, equipped, and trained by the federal government, but available to the states except during war. To ensure stability after the Great War, Scales, like other adjutants general from poorer states, needed the National Guard in existence, and would brook no competition for arms or recruits. He and others saw the need for a National Guard all around them.

The Seattle General Strike of February, 1919, and the riot in Centralia on Armistice Day, 1919, incited fears of an organized effort to establish a soviet system in America. Officials in many states believed they were witnessing the beginnings of a Bolshevik Revolution in America. Faced with domestic insurrection, riots, and racial violence, Scales, and most other adjutants general, wanted the federal government and state governments to begin to re-create the National Guard in earnest for traditional militia functions. This would give adjutants general a federally equipped, armed, inspected, and trained force for maintaining order. In addition, the federal government would provide drill and summer encampment pay for National Guard, while none was provided for home guards.

However, poorer states were not only ones frustrated by the departure of the National Guard. The state with the largest National Guard, New York, found its experience with militia during the war unsatisfying. The New York Guard organized on August 3, 1917, as a division to comply with provisions of state constitution.[51] Many smaller communities that did not have a National Guard unit in peacetime created New York Guard companies, but these tended to disband after the initial enthusiasm waned and strength could not be maintained.[52] Even in the units that maintained their strength through the war, most men were eager to get out after the Armistice. Commenting on his wartime force, the adjutant general claimed that "[a]t no time was the New York Guard properly armed, uniformed and equipped."[53] Still, as late as January 1, 1919, the New York Guard mustered about twenty-two thousand men in all ranks, but the adjutant general believed that only about five thousand could actually be called on for service because of the lack of equipment, training, and arms.

The postwar period would see major changes in the New York Guard. A new adjutant general took office on January 13, 1919, and soon began to reorganize the state's military forces.[54] To command the New York Guard, the adjutant general brought back Maj. Gen. John F. O'Ryan, former Commanding General of the Twenty-seventh Division, the National Guard division from New York in the Army of the United States. He had been discharged following the end of the war, and he and his staff took command of the New York Guard on March 31, 1919. His mission was to reform the New York Guard and ready it for providing the nucleus of the postwar New York National Guard.[55]

In contrast, some of the less industrialized states weathered the war years without major problems with small state forces, or none at all. Despite its location on the Atlantic seaboard with a long and mostly vulnerable coast, the Delaware home guard of the Great War was the smallest to receive official support. "General Order Number 10" of the adjutant general's offices, as amended by "General Order Number 21," on December 21, 1917 provided for the organization of "Companies of Infantry" for state duty.[56] The adjutant general believed that one company, plus a supply detachment, should fill the state's militia needs. This represented a force of four officers and seventy-one enlisted men.

This tiny force had its home in the state's largest city, Wilmington. The adjutant general described the force as well-uniformed, armed, equipped, and instructed. The men drilled two evenings each week for at least the first year of their existence as a unit.[57] The positioning of the force, in the northern extreme of the state, indicates that the company's main function was to respond to urban unrest. Also, the use of Wilmington for a home base reflects the sparse rural settlement of most of the remainder of the state. However, the adjutant general proved correct. No situation required the use of the force. By keeping a small force that the state could properly uniform and equip, Delaware avoided many of the headaches other states encountered. By the end of 1920, little had changed with the force. The adjutant general praised the men for their faithful service and mustered out the company shortly thereafter.[58]

Maryland had a similar experience in the war. Its National Guard entered federal service for the Mexican Border in June, 1916, and returned home at the end of February, 1917.[59] Their return to civilian life proved short-lived, for less than a month later, two units entered state active duty in order to guard railroad bridges over the Susquehanna River. At the same time, the Fourth Regiment of the Maryland National Guard again entered federal service by presidential proclamation. In April, the state ordered part of the Fifth Regiment to state active duty to protect Baltimore's water supply at Lake Montebello.[60] When the United States declared war, the state's naval militia entered federal service immediately, while the Fifth Maryland also entered federal service. Gov. Emerson C. Harrington called up the First Regiment to take the Fifth's place watching the reservoir. Later, a special municipal force led by a former National Guard officer assumed the duty of protecting the water supply, relieving the First Maryland. The Maryland National Guard not in federal service began recruiting and training until it too entered federal service on July 25, 1917.[61]

At an Extraordinary Session in June, 1917, the Maryland Legislature revised the Militia Law of Maryland to provide for a Maryland State Guard of one regiment of infantry with not more than one thousand officers and men. As with most states, the new law for the State Guard made applicable to the new force most of the state's laws for the National Guard. On August 4, 1917—one day before the draft of most of the National Guard into the army—the adjutant general's office began the organization of twelve line companies, a headquarters, and a supply company by Executive Order. The Headquarters of the State Guard, supply, and six of the line companies were based in Baltimore, with the other six infantry companies scattered throughout the state.[62] This pattern appeared in many states; the largest city, or the capital, received about the half the force, while the remainder came from the less populated, more rural parts of the state. Over half of the state's population, and most of its industries, were in Baltimore, so this arrangement made sense.[63] Maryland called its new force the Second Regiment, State Guard. A former adjutant general of Maryland, Maj. Gen. Clinton L. Riggs, received a state commission that October as a colonel to head the State Guard regiment. Riggs had served in the Spanish-American War and had since been active in the National Guard.[64] Other former National Guard officers were appointed to leadership roles.[65]

The regiment expanded when the "Motor Battery of Baltimore City" applied and was accepted as a machine gun company.[66] Like many other home guard organizations, the Motor Battery, consisting of automobile owners, had formed without state guidance. The inclusion of this new company brought the Second Regiment to its peak strength of approximately eight hundred men.[67] Members drilled twice a week at the fifth regiment's armory in Baltimore, although the lack of coal made the winter drill evenings a chilly experience. In addition to armory drill, the Second Regiment attended nine-day summer training encampments in 1918 and 1919.[68] For the 1918 summer encampment, the entire regiment marched in heavy heat to the training area in Timonium—making it the largest encampment of Maryland troops. The Marylanders earned praise for orderliness and sanitation from army officers who inspected the camp. At the camp, the men heard rumors that they would soon be accepted as a unit for federal service, which most men favored, but the Maryland Council of Defense quickly put an end to that idea, as the state could ill afford to raise another regiment for state service.[69]

The State Guard received its weapons—Springfield .45 caliber rifles—in early November, 1917, and uniforms in December. The uniforms were similar to those of the army but with buttons bearing the state logo and shoulder straps and collars of olive green to make them distinctive. Later, a red star, a soon-to-be resented requirement that the War Department mandated for state uniforms that were similar to federal uniforms, was added on the sleeves of the coat and overcoats of State Guard uniforms.[70] However, at least two of the companies had no uniforms or equipment as late as March, 1918.[71] March also saw the regiment receiving its colors. Thomas W. Jenkins, a former member of the Maryland Militia and history buff, donated a "very handsome" set to the Second Regiment.[72] Training emphasized sports and marksmanship, with the state armory the scene of several athletic meets throughout the existence of the State Guard. The regiment also sent a team to compete at the National Rifle Matches at Camp Perry, Ohio, in September, 1918. The trip was not all competition, as the State Guardsmen also attended a School of Fire, for marksmanship training, at Camp Perry.[73]

In addition to training and parades, the regiment also provided a needed service for federal soldiers and sailors on leave in Baltimore. Under the direction of the regimental chaplin, the State Guard operated a program of free lodging and breakfast for federal troops. After the war's end, the regiment estimated it had boarded over five thousand federal servicemen at the state armory.[74]

The water authorities in Baltimore tried to have the State Guard provide a full-time detachment to watch the water supply, but Riggs successfully avoided this burden. As with most states, indeed most militia, the Maryland State Guard trained for deployment in an acute crisis, not for routine, long-term, sentry duty.[75] The regiment, however, passed the months of the war without providing emergency service to the state. Its only recorded services came in early 1919, after the Armistice. The first, on February 27, consisted of two infantry companies and the machine gun company going to Annapolis to assist county and Baltimore police detachment in maintaining order during the execution of a man for murder. The man to be executed, John Snowden, was a black man convicted of murdering a white woman in August, 1918. The trial and execution brought a rise in racial tensions, and the governor employed the Guardsmen in the city streets during the morning of the execution to prevent violence by either blacks or whites. The city remained practically under martial law for a couple of days. One of the companies of the State Guard remained at the armory until Snowden's burial on March 2.[76] Through a strong show of force, authorities maintained order throughout the period, and the State Guard left active duty without incident.

The second and last period of active service for the Second Regiment came that April, when two officers, and later the whole company from Salisbury on the Eastern Shore, were called to the town of Easton to assist a small army of police, deputies, and highway patrolmen in preventing a white mob from lynching Isiah Fountain, a black man on trial for raping a thirteen-year-old white girl. The suspect managed to escape from the mob and authorities but was soon recaptured. The Guardsmen impressed the crowds with their intention to maintain order, but they also found time to lighten the situation by blanket-tossing new members in public. Nevertheless, when the trial ended with a conviction, the State Guardsmen maintained a cordon around the courthouse all night, and fifty State Guardsmen with loaded rifles and fixed bayonets kept several thousand whites from lynching Fountain.[77]

By the fall of 1919, a new National Guard began organizing under orders from the governor. In his report to the General Assembly, the governor recognized the debt of gratitude the state owed its State Guard for keeping "any element of discord . . . afraid to raise its head," but nevertheless, he desired that the National Guard soon reorganize and replace the State Guard.[78] The State Guard remained in existence while the National Guard formed, with the Fifth Regiment of the new National Guard sharing its armory with the Second Regiment. Under that arrangement, the commander of the State Guard remained in control of the armory until the dissolution of the State Guard.[79]

The original date for disbandment had been set at February 1, 1920, but was later delayed until March 1. This was done in order to hold a final regimental dinner at the state armory on February 14, 1920, followed by a parade and review. Then, two years to the day after receiving its colors, the regiment retired them and passed out of existence.[80] Enlisted men were discharged, while officers transferred to the unassigned list. They were allowed to keep their uniforms as a small token of their service.[81] In summing up the record of the State Guard after its demise, the adjutant general mentioned that they provided "splendid service" to the state. But for Guardsmen, "this service did not present the glamor of service in the Army" and its service was "not fully appreciated by the people generally."[82] Appreciated or not, Maryland's State Guard of the Great War, unlike many state militia to the south, would provide an example of an adequately trained and equipped militia, which would become the national standard in World War II.

Still, the need for a wartime militia was not apparent in all states. Tennessee began the war with better state forces than many states, especially in the South, possessed. In 1915 Tennessee created a new institution of State Rangers, who operated through the adjutant general's office but were to be paid by the county when called to duty. These Rangers were a state constabulary,[83] which gave the adjutant general, and through him the governor, a means for projecting state power separate from the militia. The force was readily expandable. Each Ranger could swear in deputies and place them on duty as a sort of posse comitatus.[84]

Tennessee did not hesitate about using this new force. Its first high-profile mission came in January, 1916. The Ku Klux Klan had been burning property and intimidating people in Stewart County, between the Cumberland and Tennessee Rivers. The governor sent in the Rangers, whose presence discouraged further Klan activity, thus ending the open terror.[85]

Despite the existence of the State Rangers, the adjutant general called the need for armed force in the state "obvious" after the National Guard entered federal service. However, the lack of legislation and funds made the creation of a new organized militia impossible. So the adjutant general instead began the organization of a new National Guard regiment for state service. Tennessee already had three regiments absorbed into federal service, but the War Department announcement that no National Guard units created after August, 1917, would be taken into federal service assured the adjutant general that he would retain control of his new National Guard. Plans for the Fourth Infantry came on October 10, 1917. The individual companies of the regiment, drawn from the eastern half of the state, underwent inspection for federal recognition between December 14 and 24, 1917.[86]

Because organizing the Fourth proved relatively easy, and because no units existed in Memphis, or in western Tennessee (except Union City), the state on December 12, 1917, began the organization of the Fifth Regiment.[87] Federal inspections of the new regiment were conducted from January 19 to February 5, 1918.[88] These two new regiments, the Fourth and Fifth, received federal recognition as National Guard, and with it came federal funds and, eventually, arms. After the Armistice, enthusiasm for the new regiments could not be maintained, and they were consolidated into a single National Guard regiment in early 1919.[89]

After the creation of the new National Guard regiments, the State Rangers did not stay idle. In fact, although the new regiments seem to have performed no emergency duty during the war, the Rangers did become involved during an incident of labor unrest in Chattanooga. Chief Ranger Maj. James P. Fyffe was placed on duty at Hamilton County to assist civil officers in maintaining peace from June 8 to 14, 1917, during a street car operators' strike. He mobilized a "number" of deputies who "performed effective service." After the strike, Hamilton County officials refused to pay the possemen, and the constitutionality of the Act which had originally created the State Rangers was challenged in the Supreme Court.[90] The case dragged on after the end of the war and the reconstitution of the entire National Guard. The county must have won its case, for the State Rangers did not appear in the adjutant general's reports after 1923.

Other states went further than Tennessee in replacing their National Guard during the war with full-time civilian forces. The Chief of the Militia Bureau listed New Mexico as one of the states that created no organized militia during the Great War. This is curious, because New Mexico had only recently been raided by a hostile army. In 1916, Pancho Villa's army sacked the town of Columbus, New Mexico. In response to Villa's raid, President Wilson had sent most of the available Regular Army on an expedition into northern Mexico to destroy the Villista army. In addition, the National Guard was mobilized to patrol the border. Most of this force had only been withdrawn a few months before the entry into the Great War. Although the United States left Mexico after the political situation there settled a bit and Villa's army had ceased to be a threat, the border remained a source of potential trouble, and the army kept a force of cavalry on the border throughout the war.

Still, the presence of U.S. troops on the border did not totally alleviate the fears of New Mexico, but the state responded with a force quite different from a home guard. Because of the "disturbed conditions along the Mexican border" and the high prices of livestock, the stealing of cattle and sheep was rampant, especially in the southwest counties of New Mexico. Because the problems facing the state were seen as more of law enforcement than military in nature, the state authorized no home guard units.[91] Instead, a Mounted Police force was formed to respond to demands from stockmen and property owners. Cattlemen and the governor, along with Sanitary and Defense Committees, formulated plans for a troop of Mounted Police at a meeting in Las Vegas, New Mexico.[92] At first, the state legislature made no provisions for paying this force, but a resolution to this problem was soon found. Because the New Mexico Mounted Police also performed the duties of Mounted Inspectors of the Cattle Sanitary Board, half the pay for the force would come from the funds of the Cattle Sanitary Board, and half from the Public Defense Fund.[93]

The troop had a captain, two sergeants, fourteen regular paid privates, and several hundred Mounted Policemen who furnished their own equipment, served only when needed, and did not receive pay for their services.[94] The unpaid men signed on for service in an emergency only. A Special Committee on State Police had direction of the Mounted Police for the duration of the war emergency. Although this force resembled a state police or constabulary more than a militia, it provided the state with its closest approximation to a home guard during the Great War. The Mounted Police also helped enforce so-called "slacker laws" requiring able-bodied men to work, antigambling laws, and traffic laws. According to the chairman of the State Council of Defense, it was useful "particularly in preventing and suppressing pro-German activities and propaganda."[95]

However, most of its work was the normal duties of a state police in a sparsely populated region with no large cities. The force captured felons, worked to prevent "white slavery," bootlegging, selling whiskey to Indians, unlawful butchering, larceny of stock, and burning brands. The force also assisted in capturing deserters from the army, although police in other states did the same type of work.[96] And so New Mexico, the focus of so much army and National Guard activity only a year before the entry of the United States into the Great War, used mounted police for most militia functions. States with a more urban population than New Mexico tended to desire a new National Guard for the unrest in their cities.

To pressure the federal government into re-creating the National Guard, state governments had a powerful ally in the National Guard Association. The National Guard Association held its first regular meeting of the postwar period, after the initial planning meeting, in May, 1919, in

St. Louis, Missouri, and immediately a problem arose over the number of delegates each state could send. The association had sent to every state military department a notification for the spring meeting that called for each state to send a representative for every five hundred men in its organized militia. The problem arose because technically most states at that time had no National Guard, but some had home guard units or a State Guard. Colonel Donnelly of Missouri became particularly upset because his state had sent representatives based on the one regiment of National Guard that had been created after the draft of 1917, and had not included his state's several home guard regiments when deciding how many representatives to send. To make the problem more complex, the states had no uniform policy on what constituted a home guard. Some states had formal enlistments where others did not. A delegate from California explained that in his state, the home guard was simply individuals who "have a licence to bear arms" from the governor. The state had no enlistment or obligation from these men to do anything, but they bore the title of home guard in state laws. The delegates argued whether delegations had one vote as a group, or a vote from every representative. The constitution of the association did not help because it used the term "organized militia" in some sections, such as the section stating the purpose of the association, and the term "National Guard" in others, such as the one dealing with representation. Since the *National Defense Act of 1916* had originally prohibited the states from keeping an organized militia except the National Guard, the two terms were usually synonymous. However, the Great War had changed that. The basic disagreement of the delegates centered on whether the National Guard Association spoke for all organized militia, including State Guards and home guards, or only the National Guard.[97]

The National Guard Association had originally been formed by representatives from the militia of several states who as a group desired more federal support and uniformity among state forces. One of their victories was the prohibition in federal law of states keeping an organized militia beside the National Guard. But another victory of the association had led to its present dilemma. The association had long fought for the role of the nation's second line of defense after the Regular Army. The president by law had to draft the National Guard before calling for volunteers or conscription. These two goals of the National Guard Association were now at odds with each other. The draft of the National Guard had left most states without any organized militia, leading some to create new forces. Now the standing and role of these new forces became a threat to the future of the National Guard, and a divisive issue at the start of the first real conference of the National Guard Association after the Great War.

After a long and sometimes passionate debate, Brig. Gen. Charles I. Martin, the Adjutant General of Kansas, began to speak. His state had a large home guard that he had been reforming and using part of as a base for a new National Guard. He described the association as the organization for and of the National Guard. The credibility of the National Guard came from service in Europe, and the strength of the association, which desperately wanted to revive the National Guard, rested on the National Guard's reputation gained in France.[98] His opinion set the consensus and settled the debate. The National Guard Association would continue to be the lobbying group for the National Guard and would not support any other organized militia. The association had chosen to continue to champion militia in its role as a reserve of the army. State service, although an excellent public relations tool, was at best a secondary function of the National Guard. Wartime service with the army would continue to be its reason for being.

The association soon set about the task of ensuring that the National Guard had a role in the national military structure in the postwar era. But not all states looked to the federal government to ensure the presence of a viable militia. Rather than in the sparsely populated West, or in the rural South, home guards of the Great War reached their greatest levels of organization and employment in the Northeast. In particular, the State Guard in southern New England demonstrated the potential for state forces in months after the Armistice. Connecticut and Massachusetts were among a handful of states that felt secure enough with their state forces to adopt a wait-and-see attitude toward the eventual form the National Guard would take. The experience of several northern states with temporary militia forces differed significantly from the southern and western experience but reached its zenith in southern New England. Connecticut's force, called at first the Home Guard, and later the State Guard, proved better organized and equipped than most.[99] Connecticut first organized its Home Guard in 1917, partially drawing on the organizational experience of its two companies of the Governor's Foot Guard and its troop of the Governor's Horse Guard,[100] ancient militia organizations with mainly ceremonial and social functions that had remained independent of the National Guard.[101] Although smartly turned out, Connecticut's State Guard, as the Home Guard had been renamed in the spring of 1918, found little opportunity to prove its mettle during the war.

After a period of draw-down following the Armistice, the Connecticut State Guard began to increase its numbers in response to what the state saw as a rising threat of Bolshevism in the nation. In April, 1920—almost a year and a half after the Armistice—the state announced plans to recruit the State Guard to its full strength of fifteen thousand men. Leaders of the State Guard cited with pride a batch of papers seized by Chicago police during a raid on an International Workers of the World office that advised fellow Wobblies to "Go slow about Connecticut: they have good troops of their own there," indicating a healthy respect for the Connecticut State Guard.[102] In making this decision to rebuild its State Guard to full strength, Connecticut empathized with the federal government's "necessarily slow-foot[edness]" but believed that unrest in the nation meant the state had an immediate need for an organized militia.[103] Until the federal government established its policy for creating a new National Guard, Connecticut would use its State Guard to maintain order in crises. During its existence, the Connecticut State Guard provided the state with a competent force that allowed Connecticut to take an aggressive stance toward its perceived enemies, and not depend on federal soldiers to keep order.

Its larger neighbor to the north had a similarly well-maintained State Guard that had been created and maintained without federal support. As a result of the effectiveness of the Massachusetts State Guard during the Boston Police Strike in the fall of 1919, Governor Calvin Coolidge emerged in the national consciousness as a champion of law and order during the first Red Scare. His actions during the strike—with his statement that "there is no right to strike against the public safety"—brought Coolidge a wave of popularity that would carry him to the presidency. This was possible only because of the foresight of the previous governor, and the dedication of thousands of State Guardsmen who wished to contribute to their state while their nation was at war.

The Boston Police Strike came as residents of the commonwealth became concerned over the threat of Bolshevik influence in the cities. An earlier police strike in Montreal increased fears of disorder. A race riot in Chicago in July, 1919, had taken six regiments of the state militia to quell. Closer to home, the northern Massachusetts industrial town of Lawrence, the scene of riots before the war, again exploded in ethnic and class violence that February. Whatever the legitimacy of the Boston policemen's grievances, city leaders feared any labor trouble as the work of dangerous subversive elements.

The American Federation of Labor (AFL) strove to distance itself from radicalism. At its annual convention that summer in Atlantic City, the AFL voted against recognition of the Soviet Union. At the same convention, however, it voted to extend charters to unions of its traditional enemy—the police. The Boston police, like most police departments in the United States, had no union. What they had was their Social Club, which had none of the influence of a union. Boston policemen had served through the war years without an increase in pay while inflation ate away their buying power. By 1919, the average patrolman worked over seventy hours a week and took home less pay than a trolleyman.[104]

Police Commissioner Edwin U. Curtis held that police were officers of the city and not employees. However, during the war, Boston policemen had been classified as employees of the city in determining their draft category. The commissioner claimed that while he did not oppose a police union, he would not allow one affiliated with a national union. City leaders feared losing control over the police. For years, city and state police had been replacing militia in breaking strikes in the more industrial areas of the United States. AFL support of police unions represented an irony for organized labor. The bitterness many workers felt toward policemen would bear rotten fruit during the strike. In the heady days of August, 1919, leaders of the city's other unions pledged sympathy strikes in support of the police. This would bring the dreaded general strike and cripple Boston. But the AFL did not exercise executive control over member unions. Only the membership could vote for a strike, and when the policemen of Boston walked off the job, many Boston workers used the opportunity to settle old scores with former antagonists rather than vote for working-class solidarity.

The police strike in Boston began in early September, 1919. Rumors of the impending strike had reached the mayor's office during the first days of the month. Advertisements for volunteer policemen appeared in the city's newspapers. However, Commissioner Curtis believed he had the situation under control. He estimated that eight hundred police would remain on duty in the event of a walkout. The actual number was just over half of that.[105] Of the 1,544 men on the Boston police force, 1,117 left their posts following the afternoon role call on September 9, 1919.[106] The Parks Police supplied 100 men for the emergency in Boston, while the State Police, still in its infancy, sent an additional 60.[107] The Parks Police came under strong criticism from the commissioner for their conduct during the strike.[108] The State Police were a more efficient but relatively new institution and lacked the manpower to assume completely the police duties of the largest city in the commonwealth. The volunteer police organization, consisting mainly of middle-aged men from the upper classes of the city, was not of the size or mettle for tedious and lengthy service. Out of necessity, the governor turned to the State Guard when order began to collapse on the first night of the strike.

In the days before the policemen voted to strike, the chief quartermaster collected equipment to support the Tenth Regiment should it be needed. Originally he had planned to move the equipment to East Armory, but he later chose Commonwealth Armory so as not to cause unrest by military preparations near a busy street. The First Motor Corps and the First Troop Cavalry, holding drill in the city that evening, were ordered to stay on duty as a further precaution.[109]

Shortly before midnight after the walkout, unrest spread throughout the city. Although hardly the riot it was termed in official histories, numerous businesses and homes were broken into by groups intent on taking full advantage of the lack of police on the streets. Crowds began forming in several squares and fighting erupted. On September 10, Commissioner Curtis fulfilled the letter of the law in such situations by informing the mayor that "the usual police provisions [were] at present inadequate to preserve order and afford protection to persons and property."[110] The same day, the mayor took control of all units of the State Guard then in the city and requested that the governor mobilize more State Guard regiments for service in the city. The next day, September 11, Governor Coolidge formally took charge of the entire State Guard, which had been in the city since early morning.[111] That afternoon, the State Guard began patrolling the city, discouraging those who had hoped for a second night of open robbery.

The Massachusetts State Guard that took control of the city had changed a great deal since the Armistice. The Richardson Light Guard of Wakefield had suffered the exodus of many of its former members after the war ended and the enthusiasm of war receded. The summer encampment of 1919, held at Camp Robert Bancroft in Boxford, came after expiration of the original two-year enlistments. Unlike most other companies, Wakefield's

Richardson Light Guard entered their second summer camp with a full compliment of three officers and sixty-one men. However, many of these were green recruits; the company's numbers had plummeted to thirty-three men only a month earlier. The Concord Minute Men had a similar story, as had most of the State Guard. By the second camp most of their original members' two-year enlistments had also expired. With the war over most men did not reenlist. The company had also lost members who entered the federal forces, one of whom died in the war, but also including three who went to federal Officers Candidate School. As a result, the company included many younger men. The Minute Men entered the second State Guard camp "a small and very green company."[112] Officers tried to practice regimental drill before many of the men knew how to march with their company.

Thus in the summer of 1919, shortly before the years of drill, skirmish, and rifle practice would be put to use in the Police Strike, the efficiency of the Concord company—indeed the Massachusetts State Guard as a whole—fell to its lowest point. Throughout the war years, when the men had developed a comradery born of men of similar age and interests, the company's only function had been to escort the town's veterans and Boy Scouts in the annual Decoration Day parade. Early in the war, the company participated with the rest of the State Guard in a parade through the streets of Boston. To the Bostonians, the sight of so many older soldiers in ill-fitting uniforms may have seemed amusing, but the day would come when the people of Boston would appreciate these enthusiastic amateur soldiers, who even in their hometown were sometimes called "Mud Guards" and who had begun to feel self-conscious about their role after the war ended.

On September 10, the Richardson Light Guard of Wakefield received instructions from their regimental commander to assemble for service in Boston. By 1:30 P.M., the company formed at their armory in Wakefield square and departed for Cambridge to join the rest of the regiment. At Cambridge, one lieutenant and thirty-nine men from another company were attached to the Wakefield company, bringing its strength up to four officers and ninety-three men. The company then went by truck to the Brighton section of Boston where they occupied Police Station Number 14. Shortly after they arrived, in consultation with the local police captain, the company moved out to disperse a mob blocking Market Square and the adjacent streets. After restoring quiet to the area, the Wakefield men set up posts at strategic positions along the streets and began to enforce order in their district, which ran from Market Square to Oak Square, and thence to the Charles River. In addition, they patrolled the "more quiet residential sections adjoining Brookline."[113]

As with the men from Wakefield, the Concord Minute Men found their greatest satisfaction in service during the Boston Police Strike. Following the return of peace to Europe, the old tattered uniforms and lack of purpose created among the Concord company a longing for a quiet disbanding of the force. The alert on September 9, and call-up on the tenth, brought new life to the Minute Men. For the Concord company, service in Boston began on an ominous note: a rifle misfired and shot a police captain in the thigh. That proved to be an isolated incident, although the condition of the weapons reflected the general state of most of their equipment. The men's uniforms consisted of little more than a combination of personal clothing and surplus state-supplied stocks, much of that from old federal uniforms. Drawing on ancient traditions out of necessity, the Concord Minute Men relied on their friends and neighbors in Concord to donate the socks, gloves, and blankets that the commonwealth could not supply. Like other State Guard companies, the Concord men collected more than just material from their hometown, they also began taking in new members. The Minute Men became the military expression of their home community sent to the big city to restore order.

The old-stock Yankees of a prosperous rural community like Concord had little sympathy with the striking Irish police of the big city. The company's supply sergeant would later refer with disdain to the police who regarded themselves as "mere laborers who had a right to stop work when their trifling grievances were not immediately redressed."[114] If the police in Boston were going on strike, the Concord Minute Men were proud to find a mission.

Many State Guard companies had disbanded following the end of the war and so the strike occupied the entire organized militia forces of Massachusetts. As a precaution against an expansion of the crisis and further delay in re-creating the National Guard, the commonwealth began to re-create the old Volunteer Militia, requesting that former National Guardsmen who had served in Europe volunteer their services again. About six thousand men did, and companies and regiments began forming. By the end of September, the situation in Boston had stabilized to the point that a general strike was no longer feared and recruitment for the Volunteer Militia ended.[115]

On September 13, the adjutant general's office increased the authorized strength of each company to one hundred men. The Richardson Light Guard detailed a sergeant and clerk to the hometown of Wakefield for recruiting. The first day eight men enlisted. Many of the recruits were veterans returned from Europe, or boys still in their teens. The following day the company moved from Brighton to the armory in Cambridge. One small boy who witnessed the State Guardsmen in their role policing Boston later remembered that "the aspect of these overage and underage guardsmen was ludicrously unmilitary. They scarcely knew the manual of arms, and they still wore the laced gaiters and felt campaign hats of the Mexican Border Campaign of 1916, which had been replaced in the American Expeditionary Force by spiral puttees and overseas caps."[116] He also recalled that, nevertheless, the uniforms, rifles, and bayonets were exciting to a small boy and made every day seem like Memorial Day.[117]

The Wakefield company moved to Roxbury for a few weeks before returning to Cambridge. In mid-October, headquarters ordered the discharge of all State Guardsmen under twenty-five years of age. The Concord Minute Men "lost some of [their] most willing soldiers, whose absence . . . could not be [made] up."[118] By October 25, the situation in Boston had stabilized enough that half of the State Guardsmen were relieved from active duty and returned to their homes. From then until the final removal of all State Guard forces on December 18, units in the city underwent constant restructuring as reductions in men continued. The Richardson Light Guard, as with most companies, allowed members with pressing business to return home first. By the end of the State Guard's involvement in Boston, only men who really wanted to remain were required to do so.

By the end of December, Boston had rebuilt much of its police force, and the National Guard was recruiting. The State Guard was becoming unnecessary. During its brief existence, the State Guard provided the commonwealth with a competent force that, unlike most states, Massachusetts needed at full strength. The existence of the State Guard allowed Governor Coolidge to respond effectively to the strike using the forces of the commonwealth, without the need of federal assistance.

After the strike, the remainder of the Richardson Light Guard returned to Wakefield and the company continued armory drills into 1920, but interest waned and a new National Guard company began forming in the town. In May, the company participated in town ceremonies for the last time. Officially, the Wakefield's company "H," Massachusetts State Guard, mustered out on November 1, 1920. A new Richardson Light Guard, part of the National Guard, took its place as the town's organized militia.

The Concord Minute Men took a similar course upon their return from the strike. Drills following the strike were lightly attended. The leadership of the Minute Men spent the remaining life of the company trying to square paperwork for the commonwealth. The supply sergeant spent an "unholy time trying to account for shoes, o'ds [olive drab uniforms], overcoats, ponchos, and a single missing rifle."[119] In the end, the state dropped its demands for a full accounting, and Concord's modern company of Minute Men ceased to exist without ceremony or announcement. A few years later, the former supply sergeant wrote a brief history of the company for a reunion dinner of former members. In it, he mentions the strike as a moment in his life of which he, and his fellow former State Guardsmen, were most proud. For the strike showed that the Concord Minute Men, indeed the entire Massachusetts State Guard, "had not been mere figureheads, even when we seemed least capable of real service. When at last we were needed, we were there."[120]

The foresight of the leaders of the commonwealth in the spring of 1917, and the enthusiasm of individual State Guardsmen throughout Massachusetts, gave Governor Coolidge an option during the strike not shared by many governors. He became identified with law and order because he was able to deal with the strike using the resources of the commonwealth. The State Guard gave him the means to restore order to Boston.

Contrasting the home guards of the nation in the Great War in this manner obviously provides too simple a formula for a complex history. With little direction on home guards coming from the federal government, states created a myriad of little armies, analogous to the militia before the rise of the National Guard and the reforms of the early twentieth century. What they created reflected their own needs, fears, impulses, and resources, rather than the national war effort. The years immediately after the Armistice, before the re-creation of the National Guard, would show the wisdom of creating a well-organized militia at the state level, as well as the perils of not having a militia at a time of social and economic displacement. Urbanized states with a centralized home guard were better able to respond to crises without federal assistance, while states without such forces could only urge the federal government to re-create the National Guard as it had existed before the war. The experience of Massachusetts, Connecticut, and other states provided the example that the nation as a whole would follow in World War II. The southern experience would be the example of what to avoid. In the next war, the War Department would take a more active role to ensure that states created well-organized forces, thus relieving the federal government of the burden of providing forces for what would more properly be state functions during wartime.

For the next twenty years after the passing of the *National Defense Act of 1920*, which re-created the National Guard much as it had existed before the war, the National Guard, and the naval militia maintained in many states, again became the organized militia of the states. The other militia forces that Massachusetts, Connecticut, and Rhode Island maintained even after 1921 provided no services to their states other than ceremonial. Only on the eve of World War II would the National Guard, the War Department, and the United States as a whole begin to reconsider the problem of state military force after federal mobilization of the National Guard.

CHAPTER 4

A National Emergency

Germany's invasion of Poland on September 1, 1939, began World War II in Europe. Although few Americans at the time foresaw direct U.S. involvement, vast improvements in airplanes and German developments in the art of war increasingly blurred the distinction between the battlefront and the rear areas. In addition, the fear of internal subversion added new uncertainties. Unlike America's experience in the Great War, the new world war forced the nation to develop a more thorough plan for internal security. The National Guard, the traditional force states relied upon in times of disaster and unrest, entered federal service beginning in 1940 and remained beyond state control until after the end of the war. As a result, most states and territories created new organized militia forces, usually known as State Guards, to fill the gap created when the National Guard departed.

The United States did not mobilize for war after Germany invaded Poland, but the nation did begin taking steps unprecedented in American history toward preparedness. For the National Guard, the outbreak of war in Europe meant additional training. Since 1916, the federal government had in most years paid the National Guard for forty-eight evenings of training per year, usually by company at the local armory. In addition, during the summer, the National Guard trained in battalions or larger units for two weeks at state-owned camps. Although the Great Depression had led in the early 1930s to cutbacks, from forty-eight to thirty-six, in the number of drills for which the federal government would pay, most units reported that they still trained the full forty-eight evenings per year.[1] By 1935, the number of paid drills was returned to forty-eight per year.

On September 8, 1939, as a response to the outbreak of war in Europe, President Franklin D. Roosevelt declared a limited national emergency. In October, 1939, the number of drills allotted to the National Guard per year increased to sixty. Seven additional days of field training were also added to the training schedule. These additional twelve days of armory training took place between October 1 and March 31.[2] The added weeklong encampment took place over the winter of 1939–40, with the specific weeks that units trained staggered over several months. Few units strayed far from their armories. The chief of the National Guard Bureau in his *Annual Report* recommended that in the following years the added week be part of the summer training period, giving National Guardsmen twenty-one days straight of field training. This proposal had been authorized for 1941, meaning that the annual training for the next summer was to have National Guardsmen spend three straight weeks at camp.[3]

This extended period of annual training never came to pass as events in Europe and Asia inspired American political and military leaders to implement a far more radical training plan for the National Guard. Instead of allocating added training days, the president ordered the entire National Guard into active federal service for twelve months of full-time training. The original plan for inducting the National Guard for a year came at a meeting on the morning of May 29, 1940, between Gen. George C. Marshall, the army chief of staff, Maj. Gen. John F. Williams, the chief of the National Guard Bureau, plus other officers from the army and National Guard.[4] On July 31, 1940, President Roosevelt called on Congress for the authority to order the National Guard to active service for the year of training. This authority was soon granted, and the president issued the executive order on August 31, which stipulated that the first units would begin their yearlong tour on September 16, 1940.[5]

In the Great War, mobilization of the enormous army had begun in the summer, which allowed the War Department some breathing space as far as building suitable shelters for the large influx of men. The situation was different in the fall of 1940. Facilities did not exist to house the entire National Guard over the winter, so the activation of the National Guard occurred over a period of several months. When the original twelve-month period was only about half complete, Congress authorized the president to extend to eighteen months the term of active service for all men inducted under the earlier act.[6] For the National Guard in the summer of 1941, this meant that the earliest any unit could be returned to state control would not come until March, 1942.

For individual National Guardsmen, the call-up for a year of active federal training meant leaving family and jobs or schools for a year. In order to make the terms of the National Guard call-up more equitable with the regulations for Selective Service inductees, Guardsmen under the age of twenty-one who had enlisted without parental consent and had served less than six months in the National Guard, were discharged. Likewise, lower-ranking men with dependents or jobs deemed critical to the economy were also discharged. This removed about 51,000, or about one-fifth of the strength of the National Guard.[7] As the economy recovered from the depression, the price of labor increased, but National Guardsmen would be earning only the standard federal military pay. Had the United States not entered the war, the National Guard as a whole would have reverted to state control at the end of its training. As with the peacetime implementation of Selective Service, the activation of the National Guard also brought questions of the constitutionality of such a move. The nation was at peace and no invasion appeared imminent. Still, the War Department received authority to send National Guard units outside of the United States, although they were not to leave the Western Hemisphere except to deploy to U.S. possessions in the Pacific.[8] Despite some units deploying to the Philippines, the bulk of the National Guard entered active federal service not to man the nation's ramparts but to train for war.

The call-up of the National Guard had consequences for the states and would also raise legal and constitutional issues over the relationship between the militia and the army, and the responsibility of each during war. With the National Guard in federal service, the states again lost their principal source of responding to riots, floods, and unrest. While many states resented the temporary loss of their National Guard, the War Department had no intention of using federal troops for the normal peacetime missions of the National Guard. Soldiers were not trained for service as watchmen and would be needed in the rapidly expanding army to prepare for global war. As a result, the War Department and the Congress encouraged the states to create a new, temporary, organized militia to fill the void. While states wanted the federal government to assume as much of the financial burden for these forces as possible, the federal government sought to limit its involvement to one of general guidance. From this period of uncertainty came the birth of the State Guard of World War II.

The experiences of the states during the Great War and, more specifically, the immediate postwar period had left many state governments and the federal government dissatisfied with the state forces raised during that war. States that relied on central direction coupled with grassroots organization and enthusiasm to create their state forces fared better than states with a more decentralized or nonexistent militia. At least twenty-seven states created some type of replacement militia for their National Guard.[9] The quality of those state forces varied widely, however. While states such as Massachusetts and Connecticut created viable State Guards that served them well during the unrest in the immediate postwar period, many other states had no real state-controlled militia at all. Locally organized militia companies often remained beyond effective state control. While usually such forces were merely patriotic groups, some became a bit zealous in their enforcement of loyalty. Several states created no forces and instead depended on federal troops in case of riot, strikes, or natural disaster. Altogether, the War Department estimated that 79,000 state troops were raised during the Great War.[10]

Originally, the National Guard was supposed to remain on active federal service for one year and then return to state control. Many states therefore did not begin to create State Guards until the attack on Pearl Harbor. They sought to avoid the burden of creating a new force, which would soon be disbanded when the National Guard returned. Those states that did begin to form a State Guard tended to contain large urban areas or sea coasts. Although invasion was unlikely, states nevertheless needed a means for enforcing law when ordinary methods broke down, as well as responding to natural disasters. For state governments, the loss of their National Guard presented a real problem whether the nation entered the war or not. Many states had small or no state police forces, and the National Guard provided most of the emergency services that a myriad of state and federal agencies provide today. Without their National Guard, states could not protect vital industries and infrastructure and would be helpless in the event of a breakdown of public order.

The War Department had no intention of re-creating the United States Guards to perform internal security missions. The War Department believed that a viable organized militia on the state level would make the recreation of a force like the U.S. Guards unnecessary. Formed during the Great War in part for areas not protected by a viable home guard, the U.S. Guards contained 25,000 men organized into forty-eight battalions. This federal force had been formed with men who had been drafted and trained but for physical reasons were unqualified for overseas service. The U.S. Guards relieved deployable men from duty at some 338 industrial and strategic points around the country.[11] The U.S. Guards had been a federal force, created, equipped, and paid by the War Department, with the men in the force belonging full-time to the army. Although the men had already been trained, the need to equip, pay, and transport them for internal security missions placed an extra burden on the army—a situation the army did not want to repeat. The U.S. Guards filled the gap during the Great War to some extent, but the War Department did not want again to use soldiers and resources for internal security missions.

The crisis in 1940 differed from the Great War in another important aspect: for the first four months after the American declaration of war in April, 1917, some 100,000 men in the National Guard protected utilities, key industries, and strategic points, as well as enforced President Woodrow Wilson's policy toward enemy aliens, before being drafted en masse into the army. That period in which the National Guard would be available for such missions would not be repeated in World War II, in which General Marshall saw American involvement as inevitable. Except in the unlikely event that the eighteen-month call-up expired without another extension, or before the United States entered the war, states would not have their National Guard when war came. The War Department had no intention of sending the National Guard on interior sentry duty when war was declared. The War Department wanted to prepare to fight a global war; the states should patrol the waterworks and suppress riots.

Despite the War Department's desire to leave internal security missions to the states, most of the proposed plans for internal security involved some sort of federal force. One plan the army received came from a member of the American Legion, and called for a new military force of "Minute Men."[12] This force, to be subject to military discipline, was to include men of all creeds and colors, carefully selected for their loyalty, fitness, and ability. The plan called for the Veterans of Foreign Wars and American Legion to provide district supervisors as guides for this movement. The drafter of this plan believed that even members of those organizations should be screened because in the Great War as well as in 1940 some members might be pro-German. The "Minute Men" would only be used for defense against invasion, and would not be used for law enforcement, which could deprive the force of the support of organized labor.[13]

Although the "Minute Man" plan would be altered, its general outline came to Congress as the "Home Defense Organized Reserve" bill in February, 1941. This proposed force would form part of the reserve component of the army. It would have missions similar to a State Guard but would be a federal force. The men in it would be paid for up to eight hours a month of military training. Enlisted men in the force would still be liable for conscription into active service, although the bill said nothing about the officers in the force. The purpose of the "Home Defense Organized Reserve" would be to defend against invasion more than for disaster relief. The federal government could use the force only in a national emergency.[14] The army completely opposed the creation of this type of force. The army never wanted, and strenuously avoided, a re-creation of the situation in the Great War. With the hostility of the War Department to the "Home Defense Organized Reserve," the Military Affairs committee took no action on the bill, and it died in committee.[15]

Despite pressure for a federal militia, the War Department remained against creating one. As a legal basis for that stance, the War Department often cited the Second, Ninth, and Tenth Amendments of the Constitution, and the court case of *U.S. v. Cruikshank.*[16] The War Department insisted that local, municipal, and state governments bore the responsibility for the protection for life and property. Throughout the war years, the War Department would stress this order of responsibility for public order. Far from trying to assume too much control over civil life, in this area, at least, the War Department strove to place as much of the burden for domestic order on the states.

The states, however, had been prevented by "Section 61" of the *National Defense Act* (*NDA*) from maintaining a militia that was not a part of the National Guard. This section also stated that it did not prevent states from keeping a state police or constabulary, but these institutions, which at that time many states did not possess, were full-time civil rather than part-time military institutions. The Constitution itself prevented the states from keeping a full-time military in time of peace.[17] In the national crises from the fall of France until the attack on Pearl Harbor, the United States was technically at peace, and therefore states could not raise a new militia to replace their National Guard without a change in "Section 61" of the *NDA.*

About this time the American Legion also presented its plan to the army for internal security after the departure of the National Guard and invited the judge advocate general to comment.[18] The American Legion's original plan called for a national home defense force using the membership of the Legion as cadre. General Marshall informed the Legion that its plan was similar to the War Department's except that the War Department wanted the force at state level. He advised the Legion to contact the commanding generals of corps areas to coordinate activities.[19] Although a sound plan, the Legion's offer was unusual in that it proposed giving a large internal security mission to a private organization. Only Alabama followed the basic American Legion plan when it created the Alabama State Guard.[20] Most other states had specific clauses in their laws preventing the enlistment in body of any such group.[21]

In the summer and fall of 1940, Congress debated amending "Section 61" of the *NDA*. In hearings held that September, the Senate Military Affairs Committee heard testimony on Senate Bill 4175, which would specifically allow states to organize replacement militia for those periods when the National Guard was in federal service. Under the *National Defense Act* then in force, states had the right to maintain another militia only during wartime. The director of civilian defense, and former New York City mayor, Fiorello LaGuardia, testified against depending on state-based forces. LaGuardia said that the situation demanded the creation of efficient forces, and he believed that this would not be the case if the forces were state forces.[22] He wanted the army to activate the remainder of the fifty-six battalions of Military Police that the War Department had included in their 1939 plan for raising a four million-man army. At the time of the debate, only three such battalions had been organized. LaGuardia agreed that state and local governments had the first responsibility to protect life and property but believed that home defense forces were unprepared to assume these missions. General Marshall and Secretary of War Henry L. Stimpson rejected LaGuardia's plea, in part because Marshall feared that any preparation by the army for internal security functions would lead states to assume that the army would fulfill that mission in time of war, and that states therefore had no need to create State Guards.[23]

Also at the hearing was Col. J. M. Churchill of the General Staff, who testified to the committee as an expert on the U.S. Guards of the Great War. He said such forces would not be necessary in the future because of the War Department Mobilization Plan for the creation of fifty-four Military Police Battalions which would do the type of duties that had been assigned to the U.S. Guards during the Great War. Despite the plans to create the State Guard for internal security, the War Department still would need its own forces for War Department installations. Federal installations and industries deemed essential to the war effort were not evenly distributed throughout the nation. States with a higher concentration of these would be required to shoulder more of the burden. Militia forces were ill-suited for use as watchmen because militiamen also held full-time jobs. As such, they could not reasonably be expected to provide long-term active service. They were for acute crises. Nevertheless, the War Department had no intention of using the Military Police of the Zone of the Interior for militia functions.

Also testifying to the Senate was F. S. Ray of the Disabled Emergency Officers Retired, an organization of former reserve officers who had been wounded in the Great War, who advocated legislation that would allow disabled veterans to keep their pensions and disability pay while serving in state or home guards.[24] The War Department was well-disposed to support Ray's plea. By summer of 1940, the army chief of staff had been planning for a state-based force but hoped to use retired army officers to head them. The chief of staff believed "these state forces will be a means for trapping much of the patriotic fervor now being directed at the War Department. Their psychological effect will be great and a show of this force will have a deterrent effect on 'fifth column' ideas and subversive activities."[25] This described perfectly two of the major missions for the soon-to-be-created State Guards, especially in the early days of their existence. Creating the State Guards would channel the impulse to "do something" that surged through the nation at the start of every major war and at the same time would serve notice to potential troublemakers that society had the means to deal with disorder. The War Department, as well as state adjutants general, had been receiving a flood of ideas from groups and individuals offering services and proposing schemes to deal with their fears. The State Guard provided a role for these people. Locating the State Guard at the state level made them the responsibility and problem of governors and adjutants general rather than the War Department.

The War Plans Division did not concur with S.4175 as written because the division feared that the creation of new militia forces could lead to the dispersal of national defense efforts. The division wanted the bill changed to ensure that states would only create and keep a separate organized militia when the National Guard entered active federal service, although including periods when the nation was technically at peace. The War Plans Division also sought to limit the material assistance the federal government would issue to state forces to surplus items not needed by federal forces.[26] The War Plans Division saw the State Guard as a means of relieving some burdens from the army and had no desire to make the State Guard itself a burden on the army.

Shortly after the Senate hearings, General Marshall recommended that the name "State Guard" be used for these forces because this name "will indicate their status as soldiers within the Rules of Land Warfare. This name is preferred to the others so far proposed."[27] The term "State Guard" had been used by the home guards of Massachusetts, Rhode Island, Connecticut, and Kansas during the Great War. The term "State Guard" could be confusing, despite its obvious contrast with the term "National Guard." Although the National Guard was essentially a national force while the State Guard belonged exclusively to the states, the potential for confusion remained. The term "State Guard" had long been synonymous with the term "National Guard" in a few states such as Kentucky.[28] Nevertheless, the War Department preferred "State Guard" over all others and encouraged states to adopt the term.

Even the generic term "home guard" did not have a clear or always honorable meaning. During the Civil War, militia units called home guards were often little more than armed thugs and vigilantes who tainted the term "home guard" in the public mind. More current and with a better reputation was the British Home Guard.[29] The War Department ordered studies of the British Home Guard and combed through them to glean any useful information.[30] The Americans studied the British example, but when the American states actually created State Guards, they drew quite naturally on the experience and traditions of their departed National Guards as the primary example of an organized militia. With the first units of the National Guard scheduled to enter federal service in late 1940, the creation of State Guards needed to begin quickly.

The president signed the "State Guard Act" on October 21, 1940.[31] All sides got a little something in the Act, but the War Department was the clear winner. With the Act, states received ample authority to create new forces without fear that these forces would be taken into federal service, although the men in state forces could still be drafted as individuals. The War Department divorced itself from all responsibility for militia functions after absorbing the National Guard. In addition, the creation of these forces would not distract the War Department from organizing or equipping the rapidly expanding army. The War Department agreed to equip state forces at a manpower strength equal to half the level of the National Guard of each state in June, 1940. Originally this equipment came from old Great War items and included only limited personal items and weapons; the War Department was not going to issue enough equipment to equip each State Guardsmen as generously as it did federal soldiers.[32] States needed to post a bond and pay shipping for all federal property, and due to a nation-wide ammunition shortage, received only ten rounds per rifle.[33] States got the right to maintain forces, but they would have to scrounge at the bottom of the manpower and material barrels to create them.

The War Department planned to place administration of all matters pertaining to the State Guard under the provost marshall general, but as that office would not be organized until July 31, 1941, they were placed under the National Guard Bureau (NGB).[34] There they remained until their dissolution after the war. Through the National Guard Bureau, the War Department developed training plans for State Guards, and issued its *Regulation for State Guards.*[35] However, State Guards remained state military forces. The War Department had little authority except in regard to the care and use of federal property issued to them and instead had more of an advisory role. The only instance in which federal soldiers would conceivably take charge of State Guards would be in the event of an actual invasion of the United States. In that case, the senior federal officer would have command over all state troops in the invasion area.

Even before Congress passed the bill, some elements in the army sought to give it a broader implication. Officers within the Regular Army who opposed any state involvement in national defense hoped that under the new bill for state militia the "National Guard might be squeezed out or elevated to a Federal force supported entirely by the United States." If the National Guard could not be permanently federalized, then perhaps "if compulsory military training is to be provided, now might be the time to seek a means of disposing of the National Guard." These opponents of the National Guard hoped that after the "present crisis has passed, State forces might fall heir to armories and training facilities now at the disposal of the National Guard." This they believed would "open the road to a more proficient Army of the United States."[36]

This sentiment was yet another echo of Maj. Gen. (Bvt.) Emory Upton, a late-nineteenth-century American military reformer who had advocated a complete divorce of the states from the national military, and an indication that at least some officers within the War Department still sought to end the National Guard.[37] It also showed that at least the author of those sentiments had little understanding of the workings of state military forces. The National Guard Association had long suspected that a strong sentiment against the National Guard existed within the Regular Army. However, the NGA would not so easily be cowered. Despite the writer's sentiments, the State Guard could never be a threat to the National Guard after the "present crisis" had passed. State Guards were firmly under the control of state adjutants general, who as a group completely supported and led the National Guard as it then existed. They would allow no other force under their control to usurp the National Guard's position.

The National Guard Association was well positioned to assure that the State Guards did not permanently replace the National Guard. The Adjutants General Association (AGA), to which all state adjutants general belonged, worked hand in glove with the NGA to ensure that the State Guards did not take the National Guard's role permanently. State Guards, as all state military forces, fell under the jurisdiction of the adjutant general of each state. The men chosen by the adjutants general to lead such forces almost all had lengthy National Guard service behind them and had long been supporters of the National Guard. State Guard officers received their commissions from their states, and the states drafted the laws for the creation and use of the forces. The original leadership of the State Guard was no rogue element trying to carve out a permanent mission but usually retired National Guardsmen who wanted to give a bit more to their state during the crisis. Under the circumstances, the State Guard would be in a strictly subordinate position when the National Guard returned to state control.

Former National Guard officers were not the only source for leaders of the State Guard. Three weeks after the president signed the "State Guard Act" into law, the army announced that reserve and retired officers could accept commissions in state forces without losing their federal reserve commissions or retired status. However, service in the State Guards did not make any such officer ineligible for recall into federal service.[38] The army hoped to have as much influence on state forces as possible without the responsibility for them. Having retired army officers involved in state forces would give the army this influence. Many veterans of the Great

War would also serve in the State Guard, including the most famous American hero of that war, Alvin York, who became a colonel with the Seventh Infantry of the Tennessee State Guard.[39] Nevertheless, former members in the National Guard, many of whom had been discharged from active duty and returned to their states while the National Guard remained on active federal service, held most of the positions of leadership of the State Guard. When the National Guard entered federal service, some 3,400 enlisted men were discharged for physical disabilities, representing about one and a half percent of the total National Guard as inducted. As the chief of the National Guard Bureau pointed out, many of these were men who had long served in the National Guard and only the strenuous nature of federal training and service made their physical conditions a liability. Many of them were fit enough for the rigors that most state service entailed.[40] These men, who had been long-term supporters of the National Guard and upholders of its traditions, would be exactly the type of men sought for the new State Guards. The army later estimated that four-fifths of State Guard officers had some previous military experience. In addition, about one-third of the rank and file had previous military experience, usually in the National Guard or from the Great War.[41] With federal conscription in effect, the State Guard originally contained men too old for federal service, or men who were exempt from the draft for other reasons. They would pose no long-term threat to the National Guard's dual role.

Not all states began creating State Guards once these forces were authorized, but most at least laid the foundation for them. By the end of July, 1941, after the announcement by the War Department of the extension of federal service for the National Guard, no fewer than thirty-five states had created at least the basis for a State Guard, with a total mustered strength of 90,000 officers and men.[42] As the total number of State Guardsmen authorized for federal assistance had been set at 139,067, the force was well on the way to meeting its target strength.[43] Still, not all state institutions welcomed the creation of a new militia. In the early spring of 1941, New Hampshire, which had no signs of subversion, was rumored to be about to pass legislation for a home guard of about 450 men under the adjutant general. One company would be in Manchester, with the remainder of the force scattered throughout the state in ten platoons of forty men each. The State Police opposed the bill on the grounds that the proper solution to local defense concerns would be an increase in the strength of the state police. Some of the state's legislators opposed it because of the cost, which in any case would be minimal. The force would only be paid when called into active service of the state for an emergency. However, the frugal legislators wanted to rely on federal troops in the event of trouble because using federal troops would not drain the state treasury.[44] This kind of thinking was exactly what the War Department did not want at the state level. Fortunately for the War Department, state leaders who hoped to place the entire burden for militia functions on the federal government were in the minority, and most states began creating new forces.

States followed no one pattern, but in general, State Guards resembled the National Guard, although the men in them tended to be older. The War Department suggested that states form infantry or military police units, but admitted that the states really had the authority to form whatever they wanted.[45] In some areas local enthusiasm led to the creation of units before the ink had time to dry on the "State Guard Act." The enthusiasm of individuals allowed states to create rapidly forces never envisioned by the War Department, although maintaining these units would prove more difficult after the initial wave of enthusiasm passed. Some states included an air element to their forces using privately owned planes. Several northern states, such as New York, Ohio, Colorado, and Massachusetts, included separate companies and battalions of black State Guardsmen. Most states immediately began to request federal equipment for their forces, although the federal government was hard pressed to issue even limited numbers of outdated rifles.

With the War Department hoping that local enthusiasm would allow states to create and sustain State Guards which would relieve the army of the burden of local security, it soon became concerned over the lack of confidence the public might have in State Guards. The front page of the Sunday Comics section of the *Washington Post* for Sunday, April 13, 1941, carried *Joe Palooka,* by Ham Fisher, in which Joe, then in the army, is on a weekend "leave" when he bumps into a Mr. Charley Kerr. Mr. Kerr explains he had joined a newly formed home guard, and that the group would be drilling that evening, and asks Joe to come. The men of the home guard unit present a ridiculous appearance dressed in surplus Spanish-American War uniforms they had bought at an army surplus store and wearing civilian hats. Some of the men wear civilian clothes. Some have swords, some have rifles, and some have no arms. The whole group is undisciplined and will not listen to "Captain" Kerr. Drill breaks early when

Kerr's wife calls him home to fix the furnace. This negative image of home guards caught the attention of someone in the War Department, who questioned if this were an attempt to *"sabotage"* their efforts to have states create their own competent forces for internal security.[46]

The less than competent image of state forces brought more calls for federalization of the forces, or the creation of new federal forces for their missions. The idea of resurrecting the U.S. Guards did not die. An exchange during the summer of 1941 between a former major who had served in the U.S. Guards in 1918 and 1919 and the NGB shows that often the men in charge did not understand the issues. The former major, then a member of a local draft board, advised the National Guard Bureau that a force similar to the U.S. Guards should be reconstituted for men classed I-B, which made them ineligible for federal service, by draft boards. The NGB's response came from a lieutenant colonel, who erroneously informed the former major that the U.S. Guards had been a state-controlled force. The man at the NGB thought that by U.S. Guards, the major meant home guards. Unsatisfied by his response from the NGB, the major again wrote to explain further what the U.S. Guards had been. The NGB replied only that the War Department did not want to or plan to use federal troops for security missions within the United States. Those were state missions.[47]

Still, the push for a federally controlled force for internal security had some strong backers. Throughout 1941, the International Association of Chiefs of Police sent to the War Department several schemes for home defense forces, usually organized at the federal level.[48] The War Department would not budge from its position, even after the United States entered the war, when calls for a federalizing of the State Guards would increase dramatically.

Instead, the War Department hoped that even more areas of the nation would create their own forces. The enabling legislation for State Guards included only areas that were officially designated as a state. As a response to this oversight, Congress further amended "Section 61," of the *National Defense Act* in August, 1941, to allow Alaska, Hawaii, Puerto Rico, and the Canal Zone, to establish State Guard-type forces.[49] Alaska and Hawaii organized forces only after the United States entered the war, while the Canal Zone, with large army garrisons stationed in it, neglected to form any militia with the available civilians. Left out of this bill was the District of Columbia, which also never organized any State Guard-type forces.[50]

Although the wartime service of the State Guards pales when compared with that of federal forces, the State Guards gave states stability and an instrument to react to security concerns in the days after Pearl Harbor. By noon on December 17, 1941, the army estimated that at least 13,556 State Guardsmen were on State Active Duty protecting docks, power stations, and water supplies.[51] There they provided a calming effect to the people who served in them and for a frightened nation. The number of State Guardsmen on active service was dwarfed by the 48,000 army troops placed on sentry duty throughout the nation at the same time, but General Marshall desperately wanted those federal troops to return to their task of preparing a mass army for global war.

CHAPTER 5

The States Prepare for War

As the War Department settled on a policy for replacing the National Guard in its state role, the real work of creating new militia forces fell to state adjutants general and local militia enthusiasts. In spite of the local character of these new militia forces, federal interest in militia ensured far more homogeneity among the State Guard of World War II than home guards had enjoyed in the Great War. The main variation in state forces in World War II came in their size, but the State Guard retained regional variations. While unique problems arose in most states, and even more so in the territories, overall regional patterns did develop. States and territories bordering the Pacific Ocean, believing themselves vulnerable to attack by the Imperial Japanese Navy, created large State Guard forces that often spent long periods on active duty early in the war, while the less populated states of the inland west created tiny forces, or none at all. Many states of the Northeast did not drastically change the methods they had used to create a State Guard during the Great War. The Midwest tended to create closer reflections of their peacetime National Guard, building modest yet adequately supported State Guard forces, as did the South, proud of its military prowess and heavily dependent on militia for controlling racial strife and natural disaster.

By the summer of 1941, the National Guard Bureau listed thirty-seven states that had begun creating a State Guard.[1] The War Department agreed to support the State Guard materially at a level equal to 50 percent of the National Guard before its induction, to a total of 111,276 State Guardsmen nationally. That figure did not include any similar forces that might be raised in the territories. The thirty-seven actually planned forces exceeding federal levels of support for the entire nation, authorizing a total of 123,527 State Guardsmen, with almost 90,000 men already enrolled.[2] One of the first states to form and employ a State Guard was Alabama, but the method Alabama used to create and administer its force stood in contrast to the process in the rest of the nation. Although the method was unusual, Alabama had a viable State Guard almost as soon as its National Guard entered federal service.

While most states had in law specific prohibitions against the enrollment of any private organization in a body into the State Guard, Alabama took the opposite approach. The state gave the American Legion of Alabama the responsibility for creating and running its State Guard.[3] Governor Frank M. Dixon apparently had no qualms over using a private organization—of which he was an active member—to form his new state military. Although unique in 1940, using a private organization as militia was not new. Before the rise of the National Guard in the late nineteenth century, most militia companies were in fact private organizations that performed militia duties when called on by the state in return for legitimization of their existence.[4] Where Alabama's organized militia in World War II differed from the militia in the rest of the nation is that since the professionalization of the National Guard in 1903, private military groups had mostly ceased to play a role in state militia.[5] Governor Dixon, a disabled veteran of the Great War, used his state's American Legion as the framework of the State Guard and to provide most of its initial members. Dixon, who had been state commander of the American Legion in 1926 and 1927, was able to use his state's American Legion as part of the war effort in a manner unique for state governments in World War II. Before the National Guard left Alabama, Dixon and the Alabama branch of the American Legion began to plan for a replacement organized militia for state duties. As a result, Alabama gave significant amounts of responsibility and authority to a private organization. In return, it was able to achieve a functioning State Guard far sooner than most other states.

In the spring of 1940, several months before the federalization of the National Guard, Governor Dixon began to receive inquiries from citizens concerned over the potential loss of the National Guard. The first letter arrived in May, 1940, from a grandfather from Birmingham who believed the world situation so critical that he felt the need to offer his services for "home protection."[6] Thereafter, letters of concern or offering services continued into the fall. Most offers of service included a request for a state commission in the grade of major or above. To such requests the governor replied that he appreciated their offer but that no specific plans had yet been made. As late as April, 1941, Governor Dixon's office was replying to inquiries about a home guard by saying that the Alabama legislature had made no specific provisions for a replacement for the National Guard.[7] State laws may not have been passed, but the governor and the state military department had actually been planning a new militia for almost a year.

The War Department and American Legion had been considering a role for the American Legion in the State Guard since at least the summer of 1940. One of the army's suggestions came to the governor in an August, 1940 letter outlining S31—"Emergency Plan White."[8] The plan listed proposed training and equipment for state forces, emphasizing riot-control, as reflected in its suggestions that state forces be armed with shotguns and clubs, and receive tear-gas training. The War Department foresaw the use of state forces as one of responding to riots and civil disorder, rather than one of opposing invasions or rounding up fifth columnists. Despite the War Department's plans, at the time it had no authority to issue arms to state troops, and the army could spare no men to serve as instructors for state troops.[9] To solve the instructor shortage, Lieut. Gen. S. D. Embick, commander of the Fourth Corps Area, of which Alabama was a part, suggested to Governor Dixon that the American Legion should have "an abundance of excellent instructors" who could fill this role.[10]

Headquarters, Fourth Corps Area, pursued the idea of states using the American Legion to train the State Guard a while longer and in early September, in a memorandum to the governors of the eight states comprising the Area, forwarded a proposal from the National Commandant of the American Legion and the Fourth Corps's reply.[11] The American Legion suggested that because it had 1,060,000 members with organization and military experience in 11,700 posts nationwide, it was best suited to provide interior defense. In addition, the average age of members was forty-seven, which made them ineligible for the draft, but still young enough for state service. The national commandant offered to place the American Legion under state and army authority for use as a home guard. Fourth Corps acknowledged the American Legion's letter and praised it as being mostly in line with its own plans.[12] The American Legion's proposal differed from the War Department's plans in that the War

Department stressed state control of home defense forces. The Fourth Corps then sent copies of all letters to each governor of its area of responsibility.[13]

The American Legion would not be taken under control of the army, nor would other states use the American Legion as a base for the creation of a State Guard. Despite the War Department's tentative approval of the American Legion's offer, only Alabama employed the Legion in an official capacity. In late summer of 1940, Governor Dixon held a series of meetings with Ben M. Smith, the adjutant general, Lt. Col. Alexander M. Patch, Jr., the Alabama National Guard's senior instructor from the Regular Army, and George L. Cleere, state commander of the American Legion.[14] The purpose of the meetings was to develop plans for the rapid creation of a temporary state force should the National Guard be called into federal service. Echoing the offer from the American Legion's national headquarters, Cleere offered the services of Alabama's American Legion to the governor.[15]

As the summer progressed, Governor Dixon began informing the would-be Alabama colonels who wrote requesting state commissions that the American Legion would be in charge of creating a home guard, effectively divorcing himself from much of the burden. The first official notification of the role of the American Legion in the formation of a new organized militia came with the publication of *State Regulation Number 4* on October 1, 1940.[16] This regulation outlined the plan for the new force that was to begin forming as soon as the Alabama National Guard and Naval Militia entered federal service. The regulation defined an eligible member as "[a]ny male citizen of the State of Alabama, between the ages of 18 and 60 who is physically fit for active duty service" as long as he was not in the federal military or reserves.[17] The Alabama law made no reference to skin color; however, it went on to say that "[i]t is contemplated that members of the State Guard will be largely composed of members of the American Legion but nothing in these regulations shall be construed as preventing the commissioning or enlisting of any person not a member of the American Legion."[18] At the time, the American Legion in Alabama did not allow black veterans to join or to form separate posts.[19] As a result, the Alabama State Guard, like the Alabama National Guard of the period, was composed only of whites.

The first indication from within the American Legion that it would be playing a role in the formation of a new organized militia came in the letter of resignation tendered by Carey Robinson of Alexander City, the vice-commander of the American Legion of Alabama, wherein he mentioned his regret that he could not assist in the formation of a home guard.[20] Other members of the American Legion had no such regrets but instead eagerly awaited this new call for their services. The October, 1940, *Bulletin to Posts* from the Alabama Headquarters of the American Legion ends with the call for members to "WAKE UP—GET THE MEMBERS, BE READY TO ORGANIZE A HOME GUARD."[21] One of those waking members was Ralph H. Collins, the chairman of the Legion's Department Marksmanship Committee in Mobile, Alabama. In January, 1941, he put out his own *Bulletin* to post commanders informing them that Governor Dixon would be handing the Legion the responsibility for forming a replacement for the National Guard. Collins indicated his concern that men without prior military experience might join the State Guard. He encouraged post commanders to begin rifle training as a method of instructing nonveterans who might join and proposed holding statewide marksmanship competitions that summer.[22] Clearly he saw the State Guard mission as one of public order and repelling invasion more than disaster relief, but these were not the only potential missions for the State Guard. In June, the Alabama Department Headquarters of the American Legion instructed all department and post commanders that the American Legion could also play a role in the "Air Craft [*sic*] Warning Service."[23]

In most other states, formation of a State Guard occurred only after the National Guard entered federal service, and thus Adjutants General became central in the formation of a State Guard. The use of the American Legion allowed the Adjutant General to concentrate on other matters. For General Smith, who served throughout this period as adjutant general for Alabama, the formation of the State Guard was only a small part of his duties. Operating with a state staff whose membership changed constantly because of federal inductions, Smith had to increase the National Guard and prepare it for mobilization, create the state's Selective Service System, and prepare for civil defense.[24] With the adjutant general heavily occupied, Cleere took effective control of the State Guard from its inception. The order for the creation of the Alabama State Guard came from Cleere in November, 1940, with the target date for formation set at January 15, the tentative date when the Alabama National Guard would be inducted.[25] Cleere ordered his posts to enlist volunteers between the ages of eighteen and sixty, except those subject to the federal draft. Legion officials desired that all officers and NCOs in the new State Guard be Legionnaires, and they expected Legionnaires to fill most of the other ranks as well.[26]

The governor's decision to use the American Legion did not sit well with other veterans' organizations. State Representative Joseph N. Langan complained to the governor that the Mobile chapter of the Veterans of Foreign Wars (VFW) was miffed at the governor's decision and asked if the city of Mobile could instead have two home guard units—one American Legion and one VFW.[27] After all, he explained, the city had two National Guard armories. The governor replied that, although he was a member of both organizations, he had decided to use the American Legion because, unlike other veterans groups, it had posts throughout the state. He added that the VFW should put aside petty jealousies for the good of the state.[28]

The governor would continue to receive pleas for an official role for the VFW. Ida Lee Merchant, secretary for the Central Trades Council—affiliated with the AFL and Alabama State Federation of Labor—wrote to the governor enclosing a resolution from her group in Mobile that expressed opposition to the concept of the American Legion having the sole responsibility for the "National Home Guard" [*sic*].[29] The Council resolved that the VFW should be included in the plans for the State Guard, and specifically that Mobile's "Robert L. Bullard" Post and its commander be given a role. The letter from the council stated its belief that because of the greater foreign experience of the VFW, its members would have been "broadened mentally to the point where they would not be inclined to use the National Home Guard as a strike breaking agency."[30] Governor Dixon replied that he did not want to change or to make an exception to his Legion plan, and he reminded her that the American Legion had the broadest geographic coverage in the state.[31] He also mentioned that membership in the Legion was not a requirement for membership in the State Guard and he welcomed a "large representation of the Veterans of Foreign Wars."[32] Members of the VFW could join the State Guard, but only as individuals.

The VFW was not the only veterans' organization jealous of the role given to the American Legion. The local chapter of the Disabled American Veterans (DAV) of the World War in Russellville complained about its exclusion from the "home guard scheme."[33] Although the name of the group would imply that members were unfit for military service as a result of injuries suffered in the Great War, this was not always true. The State Guard, because it was not liable for service with the federal forces, was able to set physical standards lower than those required by the National Guard. Many veterans of the Great War bore disabilities that allowed them to perform limited duties.

The VFW was a larger organization than the DAV, but both were smaller than the American Legion. The differences in qualifications for membership were slight, but the organizations were nevertheless different. Like members of the American Legion, members of the VFW had served in the federal military during wartime but had earned a campaign medal or other decoration that would indicate service in a combat zone. Almost any wartime veteran could join the American Legion, but only those who had served overseas in a designated combat area could be in the VFW. While Governor Dixon, as a wounded combat veteran, had interests in all three veterans organizations, he did not back away from his initial decision to give the task of building a State Guard to the American Legion. Not all groups complained. The Optimist Club of Montgomery, noting the departure of the National Guard, sent Dixon a decidedly upbeat message that he could count on it in an emergency.[34]

Dixon could use the unconditional support when he had to address the next headache—providing uniforms for the State Guard. Newspapers had reported the governor's plan to create the State Guard through the American Legion, and that convict labor would be used to manufacture the uniforms.[35] Organized labor did not object to the creation of a new militia, but it had problems with the use of convict labor. The AFL claimed that because many members of the American Legion were also members of unions, union labor had a right to fabricate the uniforms.[36] In reply, the governor said that the khaki cloth had already been bought from mills operated by the Department of Corrections and the state had idle prisoners.[37] The governor would also hear from the Alabama State Federation of Labor on the same subject.[38] The federation listed several Alabama businesses that could do the work and further requested that Works Project Administration (WPA) labor not be used, except as a last resort before convict labor.[39] The Central Trades Council also weighed in with its resolution against use of convict labor.[40] Manufacture of the uniforms eventually fell to the WPA, although the state also bought some uniforms from commercial manufacturers.[41] The state later received from the federal government surplus spruce-green woolen uniforms from leftover Civilian Conservation Corps (CCC) stocks, which were issued for winter wear.[42]

Before its induction into federal service, the Alabama National Guard stood at almost 4,000 men, up from just under 3,000 in the summer of 1939.[43] The last of the Alabama National Guard entered federal service on February 10, 1941.[44] Two days later, the first unit of the Alabama State Guard was mustered and declared ready to serve the state.[45] The original plan for the State Guard called for a force of more than 2,800 men organized into seventy companies of forty men each, plus some headquarters and support units. In reality, shortages in clothing and weapons the federal government could furnish dropped the number of companies the state created to twenty-five, with the last formed by the middle of May.[46] After Pearl Harbor, the number of men authorized per company rose to sixty-five, but the rapid turnover in personnel would make this number hard to maintain.[47]

The promptness in the formation of the State Guard soon paid dividends to state authorities. While 1939 and 1940 had not seen much state use of the National Guard, 1941 saw more need. The first use of the State Guard came in late February, two days after the last National Guard unit left the state, during a jurisdictional dispute between the AFL and the Congress of Industrial Organizations (CIO) at the Utica Mills in the town of Anniston.[48] Under orders from the governor, seventy-seven men and officers of the State Guard quickly disarmed and dispersed the strikers, remaining another thirteen days to prevent a recurrence of violence. Because of the relative newness of the State Guard, many of the State Guardsmen who actually entered state service were in fact caretakers of the state's armories who had been enlisted as members of the State Guard.[49] Caretakers were the men, usually retired National Guardsmen with many years' experience, who looked after the state's armories when the National Guard entered federal service. However, many caretakers were also members of the American Legion. All caretakers were later consolidated in the State Guard in a special Service Command.[50] Soon the Legionnaires began settling down to the business of being a state militia. Although financial limits prevented the State Guard from holding a summer encampment in 1941, that fall all officers and selected enlisted men did attend a training camp on Dauphin Island.[51] Subsequent summers would see the entire State Guard spending a week of training on the island.[52]

The State Guard structure mirrored that of the American Legion to allow the implementation of the State Guard plan as quickly as possible. Mr. Cleere, state commander of the American Legion, became Colonel Cleere, commanding officer of the Alabama State Guard. Cleere, in his dual capacity, ran the State Guard, for which he received no pay except when on state active duty—as during an emergency. The Legion structure divided Alabama into three districts, each under an area commander. In the State Guard, each American Legion area commander became also a member of the staff of the state commander, with the rank of lieutenant colonel.[53] The rank came from a state commission. State law required that each area commander take active command over all units within his district when at least 50 percent of the State Guardsmen in his area were called to active state service.

A further example of the strong influence of the Legion in the State Guard was in the choice of drill nights. While the modern National Guard conducts its drills on one weekend every month, this practice did not begin until the 1960s. Prior to the change, National Guard units usually drilled for a few hours on a specific week night every week. Alabama State Guard *Regulation Number 4* specified that units would train on the same night that the local Legion post held its weekly meetings. After conducting their Legion business, members would then train as State Guard members for an hour and a half.[54] "Drill" did not mean simply marching around the drill hall; the men also received instruction and practice in skills such as marksmanship, first aid, military courtesy, and hygiene.[55] The men were not paid for drill; they were only to be paid for active state service. Not all members of the Legion became members of the State Guard, nor would all State Guardsmen be members of the American Legion, but the large overlap in membership made this arrangement practical. With only armory training that first year for most men, the maturity, experience, and organization of the Legionnaires in the State Guard allowed them to give the state some measure of protection.

Administering and providing the leadership of the State Guard was only part of the Legion's activities in World War II Alabama. Over two-thirds of the members of draft boards were also Legion members. In addition, the Legion assisted in ensuring that men registered with Selective Service.[56] The State Guard, especially in the early months of its existence, provided the men of the American Legion with their most ambitious and visible role in the state's war effort.

In July, 1941—only three weeks after Germany's invasion of the Soviet Union—the American Legion of Alabama held its annual convention. After the opening ceremonies, Commander Cleere began to speak about the State Guard program. He began his remarks by thanking the department executive committee that, among other things, ran the Legion's side of the State Guard program.[57] Then he called to the podium a committee member, Maj. Erwin Lehmann, who was also an officer in the State Guard and had attended the convention in his State Guard uniform. After a bit of coaxing by Cleere, Lehmann came forward and allowed all to see him resplendent in martial airs. The gathered Legionnaires broke into applause at the sight of one of their own again in uniform. Cleere then explained the role of the State Guard in Alabama, stressing that "these boys are not tin soldiers. They can perform services if needed."[58] In the Great War, various "home guard" type organizations around the country had been ridiculed as men playing soldier while real soldiers fought a real war in foreign lands. Cleere was anxious to avoid that image for his State Guard in this war.

After Cleere finished speaking and the display of Lehmann ended, a former Commander of the American Legion who was also the present governor of Alabama addressed the organization.[59] Governor Dixon began by telling the assembled that "[t]his organization, of course, was the organization which has long had my allegiance and the organization which I have been interested in most and have been most deeply concerned about since its formation in 1919."[60] The governor spoke a few words about preparedness, then began to address critics of the State Guard and his decision to use the American Legion to create it. The governor explained that the only democracies still standing were the United Kingdom and the United States. To Dixon, only Britain's Royal Navy stood between America and invasion. Yet the American Legion had been accused of warmongering when it sought to prepare the nation. The equipment needed for the State Guard existed in Alabama, the federal rifles for the force were already in the state, and the state needed to purchase them as soon as possible. Forty-eight states could not each take individual action or chaos would result; Alabama had to act in concert with the rest of the nation. However, when the federal government asked states to create a State Guard, Alabama was "just about the first state to put that guard in condition."[61] The governor was correct; although Alabama had blazed a trail not followed by other states in its use of the American Legion, the

Alabama State Guard was a force-in-being before many states had even developed plans for a State Guard.

Governor Dixon responded to criticism he had received for using the American Legion to create the State Guard. Some, the governor claimed, said that "they will push them into war that way," but Dixon had stood his ground.[62] When he told the convention that he would use the State Guard as he had the National Guard—for the purpose of maintaining peace—the Legionnaires broke into great applause. Themes of preparedness and law and order appealed to these older veterans of the Great War. The governor praised the speed of the Legion in organizing the State Guard. Of the 1,100 men in the force, 70 percent were Great War veterans, many of whom also had served in the National Guard after the war. He stated his confidence in his State Guard and said Alabama owed apologies to no one. Dixon closed his remarks by mentioning something that everyone present already knew—that he himself had lost a leg in the Great War, and therefore understood the horrors of war firsthand. He left shortly after his speech and did not watch the State Guard as it paraded in uniform for the convention later in the day.[63]

After the governor finished his remarks, Cleere thanked him for giving to the Legion the mission of creating the State Guard. As the labor unrest in Anniston the past February had shown, that mission would include more than parading and participating in bond-drives. During the fall of 1941, before the United States entered the war, the Alabama State Guard would respond to three more incidents of labor unrest, each of which required at least 100 State Guardsmen, and on one occasion as many as 212.[64]

The second incident occurred in September, 1941, when a CIO strike at the Republic Steel Company in Gadsden became too unruly for the small Gadsden police force. On September 13, the two State Guard companies from Gadsden arrived at the plant and quickly removed the strikers from plant property. The Guardsmen then prevented workers from reoccupying the plant. They were augmented the next day when a company from Birmingham joined them. All State Guardsmen were withdrawn a week later.[65]

Labor difficulties would involve the State Guard two more times in the period before the attack on Pearl Harbor. Both incidents involved the Tennessee Coal, Iron, & Railroad Company in Birmingham, first in September when three companies and two detachments responded to unrest related to an attempted strike and a dispute between the AFL and the CIO.

Members of the State Guard performed sentry duties at the plant's gates, protecting nonstriking workers as they entered and left the plant, while the majority of the State Guardsmen were posted near the plant in order to respond quickly if needed. By the end of the month, after four days of duty, the Guardsmen were released. They returned to the city six weeks later for a nine-day tour in response to a new outbreak of unrest. As before, most of the Guardsmen were held in ready near the plant but not actually deployed. The Guardsmen remained for nine days until the issues between labor and management were settled.[66]

After the creation of the Alabama State Guard, sustaining the force became difficult. Legionnaires had originally filled its officer and enlisted ranks, but after the immediate dangers perceived at the start of the war subsided, interest waned. Legionnaires continued to fill the officer and NCO ranks, but the rank and file held less appeal for forty-five-year-old men. Younger men who had not yet been inducted into federal service began to take their places. Still, the leadership and framework of the State Guard remained with the American Legion. The Legion allowed Governor Dixon to create the rudiments of a viable force earlier than other states. His method brought criticism, but it worked for Alabama.[67]

The experience of Mississippi was in sharp contrast to that of Alabama, and more typical of states that began creating a State Guard before the nation entered the war. Mississippi began its State Guard in the period of national crisis before the attack on Pearl Harbor and was unsure of the exact role of the State Guard. The National Guard was supposed to spend only a year training, later extended to eighteen months, and then return to state control. Had the United States not entered the war, State Guard forces would have had to have been dismantled when the National Guard returned. States feared investing large amounts of money and effort in a force that might be disbanded before it became viable.

Mississippi would not use its State Guard until March, 1944, although like Alabama, Mississippi began building its force before Pearl Harbor.[68] Mississippi wanted to avoid repeating the mistakes made during the Great War raising its militia, and created in World War II a reasonably well-organized State Guard about half the size of its prewar National Guard, although it took far longer for Mississippi to field a viable force than Alabama. The process of building a new militia in Mississippi began a few months after the National Guard entered federal service for its year of training. By November 25, 1940, the entire Mississippi National Guard of

264 officers and 3,417 enlisted men had entered federal active duty for its year of training.[69] The State Guard came into existence by an executive order of the governor in February, 1941, with recruiting begun immediately.[70] By the next month, the force mustered only sixty-one men. In April, the state military department began organizing the recruits into an infantry regiment.[71] Mississippi considered its State Guard to be part of its National Guard, although at the federal level, the two were separate entities with the State Guard not liable to be drafted as a unit into federal service. With the regular Mississippi National Guard in federal service, the total state military force consisted of the embryonic State Guard regiment with 875 men, and the adjutant general's staff of nine.[72] Not until June 30, 1943, did the acting adjutant general, Brig. Gen. Ralph Hays, report that recruiting for a second regiment was nearing completion.[73] With both regiments, a total of thirty-seven units had been "stationed at strategic points around Mississippi," meaning that "no point in the state is further than 50 miles from one of these units."[74]

All Mississippi State Guard officers were white men, but they came from a variety of backgrounds. They tended to be a decade or more older than their equivalents in the army.[75] One first lieutenant was in his mid-fifties in 1944, having first enlisted in a military unit in 1907. Most other officers had previous enlisted experience in either the National Guard or the army, and many had served in the Great War. Most men who sought commissions in the State Guard included in their applications letters from community leaders attesting to their sober behavior and trustworthiness. The adjutant general wanted mature men with the respect of their communities as the officers for his State Guard. One major in the State Guard had served as an intelligence captain in Army Air Forces in the Pacific until his discharge for medical reasons in 1944.[76]

The state expected State Guardsmen, like most of their counterparts throughout the nation, to train without pay. However, state law provided that while on state active duty, officers were to draw pay equivalent to officers in the army. Enlisted men were to receive pay at twice the daily rate of equivalent ranks in the army. Payment came from the state—the standard practice for National Guardsmen on state active duty prior to the war.[77] The first state appropriations proved inadequate for the needs of the then-forming First Regiment. Instead, county and municipal governments took the lead in providing uniforms and funds for their local units.[78] Uniforms, although similar in appearance to army uniforms, were required to have distinctive insignia that would identify them as state forces.[79]

The situation regarding a replacement militia remained fluid throughout the nation, and state military departments freely exchanged ideas on how best to create a State Guard. Georgia had sent to other southern states an outline of its plan for the *Organization of the State Defense Corps* in August, 1940, long before it actually created a force.[80] The purpose of the Georgia State Defense Corps would be "(1) to assist in the preservation of peace and good order; (2) for the protection of vital installations and public works; (3) for the supervision of activities of aliens within the state." Members of the force, which included a Women's Auxiliary, were to be white and thirty-five years old or older. The age limit was set so as not to interfere with federal conscription, which had begun in summer 1940. Company commanders were to be at least forty-five years of age. Nominees for commander were to have the approval of the local post commander of the American Legion, among other leaders in the community.[81] Georgia's plan differed little from what the majority of states was doing, but nothing indicates that its plan was a model for other states. Instead, the similarity of state plans indicates the common experience of states with their National Guard, the suggestions of the War Department, and the sharing of ideas among states during this early, uncertain period of the State Guard of World War II. Eventually, most states approached the task of creating a State Guard through a centralized method, similar to that followed by Mississippi.

Despite the initiative shown by state military departments, local enthusiasm often ran ahead of state governments. This trend was most pronounced in the Northeast—the area that had given birth to the National Guard and had created the most organized home guard structures of the Great War. In the Northeast, the threat of war in Asia and especially Europe led to the spontaneous formation of militia groups at the town level, which in many cases would only later be absorbed into the State Guard structure. One of these types of forces enjoyed some free publicity in the November 17, 1940, edition of the *Philadelphia Inquirer.* A photo story entitled "Prominent Socialites Drill for Home Defense" showed the "Pennsylvania Volunteer Mobile Defense" of Montgomery County, a mounted and motorized militia troop, performing its tactical training at the estate of one of its members. The group consisted of forty members, described as the "sons of elite society," who lived in the vicinity of Ambler and Broad

Axe, a few miles north of Philadelphia. The cavalry troop trained every Sunday and Monday, with the cooperation and approval of the Pennsylvania National Guard and local police. The men wore dark cavalry uniforms complete with campaign hats bearing cords and crossed saber emblems, riding boots, and riding breeches. The photos showed the men eating and jumping their horses, as well as a moment from tactical training—a mounted group from the attacking force saluting a retired army major who was with the defending force. Another photo showed a militiaman holding the unit colors, which consisted of a keystone, the symbol of Pennsylvania, imposed over crossed sabers, the symbol of cavalry.[82] Noblesse oblige, state pride, and a desire for adventure combined in one of the first home guard forces of World War II and one smartly attired long before most would begin to receive any uniforms.

The Pennsylvania Volunteer Mobile Defense showed what the upper crust of society could do in a decentralized approach. But not all communities had enough men with their own horses to ride, or estates at which to train, or enough money to outfit themselves so sharply. A more common pattern in the Northeast would be a town government stepping into the void and leading the initial drive for a local home guard as the National Guard prepared to depart the area. Haverhill, Massachusetts, a few miles from the Atlantic on the Merrimac River, followed a pattern as old as the town itself. Town leaders began the process on March 19, 1941, by appointing Edward M. Evan, a local resident who had served as an officer in the Great War, as provisional captain and recruiting officer of the town's new militia company. Five days later the first twenty-nine men signed on for three years. Most of these initial recruits had been in the National Guard but were ineligible for induction because of age or dependents. The company had neither equipment nor uniforms at its first drill that night in the armory, but from then on, the unit drilled each Monday night, while NCOs also trained on Tuesday nights. By the end of May, 1941, the company felt competent enough to apply for recognition in the State Guard. After the acceptance inspection by state military officials, Haverhill's militia became Company I, Twenty-fourth Infantry, Massachusetts State Guard.[83] Uniforms were received in August, 1941, but when ordered to state duty a few days after Pearl Harbor to protect railroad bridges, the men turned out in civilian clothing and campaign hats because their summer uniforms were far too lightweight for cold December nights in Massachusetts.

A similar pattern would occur throughout Massachusetts, as well as other states, although not all communities that created a militia did so for the same reasons. In January, 1941, when the local National Guard unit left the central Massachusetts town of Gardner for its year of training, Representative Fred A. Blake explained to local veterans his belief that when the National Guard returned, communities that created a replacement would be more likely to have a National Guard company based in the town. Gardner wanted its National Guard company to return to the town after its release from federal service, and so a month earlier had taken the first steps to create a home guard to replace its National Guard company. In early February, 1941, Fred G. Kegler, a past commander of the Gardner post of the American Legion, was appointed recruiting officer and custodian of the armory, with the mission to enlist a home guard. By April, the town's company numbered fifty-four men and two officers when it was inspected and accepted as Company E, Twenty-first Infantry, Massachusetts State Guard.[84]

This pattern of local initiative for a home guard that would later be absorbed into the State Guard had been used in Massachusetts during the Great War but also reflected far older traditions of raising Volunteer regiments. Massachusetts, which had been at the forefront of the State Guard movement in the Great War, did not drastically modify its method for creating wartime militia in World War II. Not all states of the Northeast followed their methods from the Great War during World War II, but as a whole, the region had strong traditions of local militia initiative and so did not need to be prodded by the National Guard Bureau, or even state military departments, to create a militia to replace the National Guard.[85]

A more radical plan for replacing the National Guard, completely dependent on local government, was mailed to several state governors from a former governor of Oklahoma, William H. Murray, who saw the nation's main threat coming from President Roosevelt. To Murray, President Roosevelt had federalized the National Guard during peacetime as a prelude to a coup. Murray believed that Roosevelt would soon order American communists to ferment unrest. Without a National Guard under state control, governors would need to call on federal assistance. Roosevelt would use the states' inability to maintain order as an excuse to make himself dictator. Murray proposed to create a posse comitatus. He believed his plan harkened back to "Ancient Briton" and he called on sheriffs to enroll men on the county level for a posse that the governor could also call up for state emergencies.[86]

As far-fetched as Murray's fears might have been, his proposed solution was far from unusual. During the Great War, several states, including Alabama, had responded to the induction of their National Guard with similar plans. The twenty years between the end of the home guard of the Great War and the creation of a State Guard in World War II had witnessed a continuation of the trend toward increased federal involvement and professionalization in state militia. Rather than drawing on traditions from "Ancient Briton" or even the nineteenth-century American militia, states in World War II looked mostly to their National Guard as the model for their State Guard.

Another model for the State Guard came from across the Atlantic in the wake of Germany's invasion of Poland in September, 1939. The British Home Guard existed to fight against invaders, but the images from newsreels and magazines of the grassroots efforts of middle-aged Britons to defend their nation struck a responsive chord in the United States and would continue to influence the way Americans saw their own State Guard.

With a realistic threat of invasion in 1940 and 1941, the training schedule of the British Home Guard was far more intense than anything experienced by their American counterparts. Home Guards were expected to train forty-eight hours each month, whereas Americans might train as little as eight hours a month. The lack of effective weapons issued to Home Guards caused supporters to question whether the War Office truly expected the Home Guard to provide a creditable opposition to a German landing. Although most Britons realized that British industry and finances were hard pressed to arm all active forces, some suspected that the Home Guard's role had more to do with channelling enthusiasm and creating propaganda than providing real security.[87]

The greatest similarity between the British Home Guard and the American State Guard was their local support and the grassroots nature of their creation. The difference between the American and British experience with local militia defense in World War II lay mainly in the American federal system. State governments were able to support and lobby for their State Guards in a way unimaginable in Britain. In addition, British Home Guards received no pay for services, while most American states paid their State Guards for state active duty, and sometimes for field training. Lastly, unlike the State Guard, the Home Guard formed entirely to oppose an enemy invasion and avoided whenever possible missions related to civil unrest or natural disaster. As with their American counterpart, the British Home Guard contained mostly overage men and men too young for induction into the active army. Unlike their American counterparts, the British Home Guard formed part of the military forces of the nation. Nevertheless, as with the American State Guard, the British Home Guard constantly lost men to the active army. In later years, the British government resorted to conscription to fill the Home Guard, although civilian courts seldom punished men who did not appear for drill.[88]

Prime Minister Winston Churchill saw the Home Guard as an example of Britain's resolve. The Home Guard exemplified the "nation in arms" ideal. The existence of the Home Guard sent a signal to both the United States and Germany that the British would indeed fight German invaders on the beaches, fields, and streets. For this reason, Churchill and many members of Parliament supported the Home Guard in its quest for a combat role. Home Guards relished the idea of fighting the Germans and did not quietly accept War Office plans for using the Home Guard for protecting bridges or simply reporting the presence of Germans. The question over guerrilla warfare or static defense was never completely settled. As with their American counterparts, the British Home Guards never had to grapple with German invaders. They did, however, become heavily involved in the less glamorous but nevertheless necessary work of civil defense and manning antiaircraft weapons.

Americans on the state and federal level took an interest in the British Home Guard from its earliest days. In early 1941, the Massachusetts Committee of Public Safety sent two representatives to England to assess the British Home Guards and report on the feasibility that such forces might meet the needs of Massachusetts. According to British reports, the Bay Staters were favorably impressed at the effort devoted to the Home Guard.[89] However, the greatest interest came from the War Department, which carefully studied the Home Guard for lessons on employment and training. It also sent investigators to Britain and maintained an interest in the British experience throughout the war. Colonel Elbridge Colby, a National Guard officer and historian who would later become a defender of the National Guard, submitted around the time the United States entered the war a lengthy report on the British Home Guard to the War Department. In it, he stressed the potential role of the British Home Guard in opposing enemy invasion and their static, die-in-place nature. He also drew attention to the strong links between the British Home Guard and the British Army. The last part of his report made recommendations on the status of American State Guard. Although Colonel Colby was dedicated to the idea of state participation in the national war effort through the National Guard, he was so favorably impressed with the British example that he made tentative recommendations on bringing the American State Guard under federal control, at least when on active service.[90]

The War Department maintained its interest in the British Home Guard throughout the war.[91] Military Intelligence included confidential reports on Home Guard methods and subjects for training.[92] Confidential reports from October, 1942, covered the role of the Home Guard in the defense planning for London. The War Department also studied training exercises held in June, August, and October, 1943. What seemed to have intrigued the American observers most was the inclusion of the British Home Guard as part of the Crown's military forces, and that the British Regular Army maintained an intimate relationship with the Home Guard, with Regulars detailed to assist in training and joint exercises held. The Americans concluded that the British Home Guard was a competent force, that the men in it were above average in intelligence, and that despite their limited training time, they performed well, especially in the role of coast artillery.[93] However, the War Department, though intrigued by the inclusion of the British Home Guards into the national military structure, never attempted to bring the American State Guard into any standing similar to that of their British cousins and instead fought hard against the idea.

Still, federal reluctance did not prevent well-meaning citizens from using the British Home Guard as an example of what the American State Guard should become. J. K. Howard of Boston wrote to Maj. Gen. William Bryden of the Office of the Chief of Staff suggesting that the State Guard be used as a base for a nationalized American Home Guard, similar to the one in Britain. As with such suggestions from before as well as after the entry of the United States into the war, Howard assumed that a federalized force would receive federal equipment at levels higher than the War Department was prepared to give purely state forces. He believed that the men in the State Guard were excellent material for such an American version of the Home Guard, but a lack of equipment and training facilities prevented them from reaching their full potential.[94]

As late as the fall of 1941, the American Legion sent a delegation to England to study the British Home Guard.[95] But despite continuing interest in the British experience with local defense forces, the British Home Guard would never be more than a minor influence on the State Guard. Americans relied most heavily on their experience with the National Guard, the various State Guard from the Great War, or in a few states, older militia traditions when they began creating a new militia to replace the National Guard.

By the beginning of December, 1941, virtually none of the National Guard remained under state control, while at least thirty-six states and one of the territories had taken steps toward creating a State Guard.[96] Quality and quantity of the State Guard varied widely, with some states having uniformed and trained units already providing state service, while others simply had laid the legal groundwork for a State Guard. The War Department had given advice, old rifles, and a few CCC uniforms but maintained its commitment to relying on state military departments to raise forces for traditional militia functions while the National Guard remained in federal service. With the National Guard's period of active service for training extended at least six months, State Guardsmen could look toward remaining embodied at least until the spring of 1942, when the National Guard was expected to begin returning to state control.

CHAPTER 6

America Enters World War II

Despite a year and a half of preparation for war, the actual entry of the United States into the war occasioned outbreaks of fear throughout the nation. General Marshall's efforts to the contrary, urgently needed soldiers were siphoned away from training and combat units in the army to provide sentries at thousands of factories, bridges, dams, and beaches. Fearing that this drain on manpower might occur, General Marshall had reluctantly agreed in the fall of 1941 to provide military police training to one infantry regiment in each of the eight National Guard divisions due to be triangularized.[1] That military police training had barely begun by December 7, 1941, and the demand for sentries far exceeded the ability of eight regiments to fill. After the attack, requests for details of soldiers came to the War Department from industries, communities, states, and departments of the federal government.[2] Every part of the nation seemed to have a vulnerable dam, a vital rail link, or a factory on which the whole American war effort depended. Corps Area commanders began taking soldiers from combat units and assigning them to sentry duty. Ten days after Pearl Harbor, the army had more than 48,000 men occupied with internal sentry duties. The War Department estimated that another 13,500 State Guardsmen were also providing similar services to their states, most likely far below the actual number of members of organized militia forces on duty.[3] The State Guard in many areas was still in its infancy, and proper control over locally created militia units was simply not always possible. As with the Great War, indeed throughout American history, the outbreak of war brought hurried meetings in thousands of communities that resulted in the formation of patrols of volunteers. During the uneasy first days of the war, these patrols, often with their own weapons, or even no weapons, clothed in their civilian attire augmented perhaps only by an arm band, patrolled towns and villages.

Much of this activity by the soldiers and militiamen was, in retrospect, wasted effort. No Japanese commandos were loose in the heartland ready to wreak havoc, no well-organized groups of German Fifth Columnists prowled the nights looking for targets of sabotage. On the other hand, State Guardsmen served a vital function in the early, uncertain days of the war, showing the public that society remained organized and was ready to defend the nation. The sight of so many soldiers and State Guardsmen acted as a national security blanket by providing readily apparent evidence that the nation was taking action. Many members of the State Guard were Great War veterans and had been training at least once a week for months before the beginning of the war. Patrolling town streets, power stations, and water supplies gave State Guardsmen a role to play in their communities with the nation again at war.

As comforting as the sight of armed men in uniform protecting vulnerable points throughout the land was, such activity was not about to bring the war to a speedy or victorious end. As soon as the initial shock of Pearl Harbor began to recede, General Marshall began reassembling his armies for their task of defeating Japan, Germany, and Italy. State Guardsmen began to stand down in most areas, although their training and recruitment took on a new sense of urgency. To replace soldiers and State Guardsmen, General Marshall stressed that the proper man for protecting factories and other vulnerable points was the civilian watchman. The army itself had long hired civilian watchmen for War Department buildings.[4] For Marshall, permanent watchman was simply not a proper role for a soldier, federal or state. Part of the reason rested on simple military principles: an army scattered throughout the nation in small detachments ceases to be an army at all. Marshall based his arguments against using soldiers as watchmen on the nature of a soldier, who tended to be a young man, prone to impulsive behavior, and not trained to function on his own. In addition, his uniform made it impossible for him to uncover secret plots. General Marshall did increase the number of military police units in the nation, but these, he stressed, were for use as an emergency reserve to be used only when private and local security methods failed and the State Guard could not handle the problem.

The role of state military forces remained ambiguous throughout the war. Basically, they existed to fill the state missions of the National Guard in peacetime, but not all states followed this simple formula, and most members of state forces were unsure of their role. Following Pearl Harbor, many states called their State Guard to state active duty, but the cost to state treasuries and to State Guardsmen themselves soon led to an abandonment of this role. Most state governments and State Guardsmen soon came to realize that state militia existed to respond to emergencies that might befall a state, although the nature of those emergencies remained undefined. For the men in the State Guard, usually middle-aged men with families and full-time jobs, prolonged full-time state service was a hardship. State Guardsmen took pride in their role as the final instrument of order in the states during the crisis. But if the State Guardsmen took pride in the role they played in protecting their communities and states, one of the bases of that pride was about to be pulled from their hands.

Only months after the war began, the federal government began to recall federally owned weapons in the hands of state forces. Since the fall of 1940, the War Department had been providing arms for state forces when the states requested them, to a level of 50 percent of each state's National Guard in June, 1940. State Guard units received older federal arms, usually the Enfield 1917, a .30 caliber rifle, along with a sling, bayonet, and scabbard for each rifle.[5] After the United States entered the war, the federal government became hard-pressed to satisfy the seemingly endless demands for weapons from federal forces as well as from allied forces. Given the acute shortages after Pearl Harbor, arming State Guards fell to a very low priority. With the need for arms reaching a critical point, the War Department in February, 1942, began to eye the thousands of federal rifles in the hands of the State Guard. Initially General Marshall opposed recalling the rifles. He believed that a withdrawal would effectively abolish the State Guard, for an unarmed militia was not a militia at all. He also discounted the idea that weapons with State Guardsmen or units not on active duty were wasted.[6] Nevertheless, the idea of a recall of the federal rifles slowly gained momentum. On March 19, 1942, the War Department ordered that in the future, the semimonthly reports that state military departments sent to the War Department would include the status of all federal rifles, as well as all pistols and machine guns in the hands of the State Guard. The reason was that, even though those weapons were not controlled items, "their status of supply is so critical."[7] The problem of arming the State Guard grew far worse in the spring of 1942 with the recall of the rifles. Many states howled in protest, claiming that such a move would destroy the State Guard. Some states attempted to purchase their own weapons on the open market, but the War Department looked with disfavor upon using materials or arms factories to make weapons for the states when federal and allied needs remained acute.

By June, in order to provide replacement weapons for the recalled rifles, the adjutant general of the Army directed the National Guard Bureau to begin issuing federally owned shotguns to the state forces. The army owned the shotguns for domestic disturbances, but members of the State Guard bitterly resented the effective replacement of the rifles with shotguns. Shotguns by their nature were more effective against civil unrest than in a combat role. In addition, shotguns called to mind images of militiamen armed with sporting pieces rather than military weapons. The firepower of the State Guard was later increased by the issuance of enough Thompson submachine guns to allow each company to have a few.[8] Although State Guardsmen were sometimes bitter over the loss of the rifles, they accepted the new weapons as better than nothing and began training with the shotguns.

The War Department balked at issuing to the State Guard additional weapons beyond shotguns and submachine guns. In the first year of American involvement in the war, every request for weapons—and many requests arrived in Washington—had to be weighed against the potential threat each state faced. The New York Guard in January, 1942, wanted to organize a coast artillery (antiaircraft) brigade for the New York City area. State military leaders requested that the National Guard Bureau consider the possibility of persuading the War Department to issue the equipment to New York for such a force. The Bureau replied that given the levels of availability of such equipment and the need for it by federal forces, New York should not count on receiving any.[9] This answer was in line with long established War Department policy that State Guards existed for internal security missions; federal forces would defend the nation against enemy attacks. Instead of accepting this division of state and federal military responsibilities, many in the federal and state governments, as well as some State Guardsmen, began to push anew for the absorption of the State Guard into the federal military.

Although the issue of state rather than federal control of militia forces seemed to have been settled in the prewar period, in the first year of the war several calls for bringing the State Guard under federal control reached the War Department. On January 5, 1942, the War Department held informal discussions on federalizing the State Guard. The discussions went so far as to produce a draft of the legislation needed to implement a federalization. These discussions, held under the auspices of the office of the judge advocate general, favored federalizing the State Guards. The members of the group discussed the legal aspects of militia under national and international law. They believed that a federalization of the State Guard would be the means to ensure unity of command between it and federal forces should an invasion occur, as well as ensure unity of equipment and organization. However, they did not look toward moving the State Guard beyond traditional militia functions.[10] This discussion was all for naught, as General Marshall had long since become committed to state control of militia forces, and the discussions of the group at the judge advocate general's office would go no further. But this would not be the end of the impulse to federalize the State Guard.

In Congress, new laws were proposed that would reorganize the State Guard as a hybrid state-federal force. The proposed law would have transformed the State Guard into a force more akin to the peacetime National Guard, with states assuming a slightly heavier economic burden than they had in recent years with the National Guard. The main advantage of the proposed law for the State Guardsmen themselves was that they would receive pay for training, with half the cost borne by the federal government. However, such proposed laws never got far, and the State Guard continued to train without pay in most states.[11] Rather than have the federal government assume any more responsibility for the State Guard, Secretary of War Henry L. Stimson believed that state-controlled militia forces, backed in an emergency by the fifty-four military police battalions, should be enough to cope with any unforeseen problems. To him, the State Guard as constituted would be sufficient. He believed that the experience from the Great War showed that home guards, when states organized such forces, rendered excellent service.[12] The War Department, with the explicit support of Secretary Stimson and General Marshall, was not about to budge on the issue of assuming control of militia.

Throughout the spring of 1942, the fear of Japanese attacks on the continental United States remained strong. In May, 1942, a group calling itself the "US Guard Committee" became concerned with a forty-mile-wide fog bank that hovered off the coast of California, fearing that it would be used to hide Japanese ships for commando raids or fire attacks on coastal cities and forests. The committee quoted the governors of Arizona, California, Mississippi, North Dakota, Oregon, as well as a few others, in support of a federalization of the State Guard, or, more to the purpose of the committee, the re-creation of the U.S. Guards that had existed in the Great War.[13] Most of the statements from the governors were taken out of context or with key parts left out. Most governors wanted more federal arms, equipment, and money for State Guard, but few if any desired to lose control of yet another organized militia that they had painstakingly built up.

Still, a small minority of men in the State Guard did desire federalization, although this may have been more in the hopes of qualifying for federal military benefits than in the interest of military efficiency. Many other State Guardsmen who supported federal control also assumed such a move would bring vast increases in federal equipment long sought by the State Guard. One state soldier who wrote directly to General Marshall cited as his reasons for federalization the lack of transportation available to state forces besides private automobiles, the lack of ammunition or a common Table of Organization and Equipment (TOE) among states, and his largest problem, the huge turnover in men.[14] Most likely, he assumed that federal control would mean not only that equipment would become available but that State Guardsmen would be exempt from federal conscription. Such exemption was never to be, and not only would State Guard units constantly hemorrhage men to the federal military, but in January, 1942, when many State Guardsmen were on active state service, the federal government began asking governors to discharge all federal employees from state military forces because their absences were hampering the work of the government.[15]

Some State Guard units contained many federal employees. In the winter of 1942, legislation had been proposed allowing federal employees to take leave for State Guard training or service.[16] Under the proposed law, federal employees would suffer no loss of pay or annual leave for time lost to State Guard service. The law did not pass. As always, the supremacy of the federal war effort was stressed at the expense of internal security in the states.[17] The War Department believed the State Guard to be important to the national war effort, but keeping federal employees at their jobs was more important.

Related to the problem of federal employees in the State Guard, and a much greater drain on membership in the State Guard, was federal conscription. Although the War Department had made clear early on that membership in the State Guard did not make a man ineligible for the draft, it had not yet ruled on whether states could prevent State Guardsmen from resigning in order to join the federal forces. Often, a man would desire to enter the federal service after getting a taste of military life in the State Guard. The adjutant general of New Jersey, on July, 1942, petitioned the army to allow states to refuse to discharge certain personnel for enlistment into federal service. New Jersey had lost so many key State Guard personnel to federal service that the adjutant general believed he should be able to prevent the discharge from state forces and the induction into federal forces of the men who provided the leadership of the State Guard. Although this proposal had been approved by the commanding general of the Second Services Command, of which New Jersey was a part, the War Department killed it almost immediately, because Marshall had no intention of allowing state forces to sap the pool of manpower available for the federal forces.[18]

Still, federal conscription added enormous difficulties to an adjutant general's ability to maintain a viable militia. Between January 1, 1942, and April 30, 1943, to cite one period, the New Jersey State Guard, which had an authorized strength of more than 2,100 men, lost 946 to the draft. Of these, 56 entered federal service as officers, and most of the remainder quickly became noncommissioned officers. In that same sixteen-month period the State Guard throughout the nation lost at least 55,000 men to the draft, of whom at least 3,000 became officers in the federal military. The rapid turnover in personnel in the State Guard, averaging over 100 percent annually in some states, meant that almost all State Guardsmen later inducted had less than a year of state service before receiving their draft notice. Officers in the State Guard averaged nine months of state service before induction, with Indiana having the longest average service with fourteen months, while South Dakota officers served only four months before their induction.[19] Enlisted men served shorter periods before their induction, with the average amount of time in a State Guard at seven months, Michigan having the lowest average at three and Massachusetts having the longest at eleven.[20] Such losses were to continue through the end of the war.

Although State Guards were in no way part of the federal military, they nevertheless came under the scrutiny of the War Department. Because they received federal property, mostly the weapons in the early years of the war, the War Department had an obligation to ensure that such property was well maintained and safeguarded. Another reason was constitutional. As militia, Congress had the power to prescribe the discipline for the State Guard and ensure that the states were enforcing that discipline.[21] However, army inspections of the State Guard often served only to stir resentment among the State Guardsmen. The inspectors, who came from the army's Administrative Service, were more accustomed to inspecting units of the Army of the United States, or of the former National Guard, where stricter standards were required. For the older, unpaid, State Guardsmen, who had no federal standing, federal inspections became a source of resentment.

Fortunately, the Chief of the National Guard Bureau understood the resentment of the State Guardsmen. A few months after Pearl Harbor, General Williams reminded Maj. Gen. John P. Smith, chief of the Administrative Services, that the State Guard was a voluntary state organization. Its morale had just received a terrific blow by the recall of the federal rifles, and General Williams was anxious not to cause any unnecessary ill feelings. He sent a confidential letter to all Corps Area Commanders telling them that several state adjutants general had recently let him know that they did not like the attitude of army officers who inspected State Guards. In the future, General Williams wanted inspections to be more of a friendly visit that was to take place during regularly scheduled drills rather than requiring State Guard units to assemble at the convenience of the inspecting army officers. While care of federal property was to be noted, aid and encouragement were to be the main purpose, not criticism. He also suggested that the army officers be called advisors rather than inspectors. General Williams, a National Guard officer himself, had sympathy with the men who volunteered their free time to serve in the State Guard.[22]

Despite the desire of General Williams to maintain morale in the State Guard, he could do nothing to prevent the much resented recall of the federal rifles. The states, however, were less than enthusiastic about returning them, possibly hoping to delay long enough for the War Department to change its mind. As late as September 14, 1942, some of the federal rifles had not been returned. In the First Service Command, Massachusetts still had 1,188, while Connecticut had 627. In the Third Service Command, Maryland had 10, and Virginia had 29. California, in the

Ninth, had the most, with 2,500 federally owned rifles. At least in Virginia, some of the rifles may have been lost, but the real reason for the delay stemmed from the very reason states had created a State Guard—they feared being left defenseless.[23] Rather than wanting to give up weapons, some states requested more. Massachusetts, in the fall of 1942, requested more submachine guns, but due to a large quantity of them being sent overseas, the War Department denied this request. The commonwealth also wanted three hundred Vickers machine guns for airport security. In its denial, the War Department stressed that, in addition to the shortages of such weapons in the federal forces, maintenance would pose a problem for state forces because of the nonstandard character of the weapons.[24]

Despite the delays in returning the federal weapons, most states had turned in the rifles by midsummer. Only Georgia and one other state had made no progress in turning in the weapons. The Georgia State Guard protested to the War Department that it had promised to replace the rifles with double-barrel pump-type shotguns, but Georgia had instead received single barrel shotguns, which it believed were inadequate. The War Department gave the Georgia State Guard until August 31 to return the rifles. Despite protestations that doing so would be impossible, Georgia managed to return the weapons on time.[25]

Despite the recall of the federal rifles, states wanted more material support from the War Department. Individually, state military departments had little chance of influencing a decision of the War Department. They needed to present a united front if more of their demands were to be met. The National Guard Association was the normal vehicle through which states lobbied for support for state forces. With the National Guard in federal service for the duration of the war, the National Guard Association for all intents and purposes ceased to function, more for practical reasons than owing to any plan. The leaders of the National Guard were busy with full-time military problems and not available for Association functions. Nevertheless, one influential group of National Guardsmen remained behind—the adjutants general of each state. The war left them in an unusual position. Since the passage of the *National Defense Act of 1916*, the National Guard had been the vehicle for state participation in the national military. Adjutants general became state officials with federal rank. The induction of the National Guard left the adjutants general as generals without their normal armies. But, as their governors' chief military advisors, and commanders of state military resources, adjutants general became, by the definition of their position, the head of State Guard in each state. Organizing and administrating these replacements for the National Guard became the primary mission for many of the adjutants general during the war.

During peacetime, adjutants general, as a group, had enormous influence over the National Guard and national policy toward state-based military forces. With the National Guard in federal service and the federal military greatly enlarged, their influence on the national level dropped. Before the war, many adjutants general had a dual assignment, as adjutant general as well as a tactical command, usually of a National Guard brigade or division. Most of the adjutants general without a tactical assignment were inducted into federal service in 1940 and 1941 and assigned as the executive of their state's Selective Service, usually at the request of their governor and the director of Selective Service. Twenty-five adjutants general entered federal service this way, while another four also became head of their state's Selective Service while remaining in their state status as adjutant general. All together, some 483 officers on state staffs entered federal service while remaining in their state staff positions during the call-up.[26]

They and their staffs were the only National Guardsmen in the nation with full-time positions who were not part of the active army. If the National Guard was to be re-created after the war on lines similar to the National Guard of the interwar period, the adjutants general needed to keep the organized power of the National Guard Association together, at least in skeletal form, throughout the war. This was done through an organization that existed more informally before the war, the Adjutants General Association. The National Guard Association could not function, but the Adjutants General Association could continue to function as a de facto cadre of the National Guard Association.

The Adjutants General Association held its first general meeting since the beginning of the war in April, 1942, in Washington, D.C., only a few days after the federal government had ordered the return of the rifles. As one would expect, this meeting focused mostly on the immediate problems facing the adjutants general—the State Guard—rather than the postwar role of the National Guard. The president of the Adjutants General Association, Brig. Gen. Charles H. Grahl of Iowa, opened the meeting with a discussion on state defense forces, but even this seemingly obvious agenda topic soon became a point of contention.[27] One member of the conference argued that they should not be discussing the State Guard because, as he read the law, the states had simply asked the federal government for some obsolete equipment for state forces and the federal government obliged. For him at least, State Guards were a local matter and not the business of the AGA. Despite his dismissal of the topic, the majority of the attendees voiced their concern over the federal government's inability to issue equipment to the states, the state's inability to buy equipment on the open market, and the general lack of available equipment.[28] The AGA would have to present a unified position to the War Department if State Guards were to get more federal support. The State Guard had a further problem when conducting training on active duty military installations. Originally, commissary officers refused to allow State Guard leaders to purchase food for state troops on the federal installations. Eventually, the War Department, recognizing the need to provide sustenance for state troops on active duty for training, allowed State Guard units to purchase the needed goods.[29]

Another issue with which the assembled generals wrestled, or at least vented their frustration, was that of state versus federal control. One idea current at the meeting was to incorporate the State Guard as a component of the National Guard. Maj. Gen. Ellard Walsh of Minnesota argued against this idea.[30] He wanted the new state forces to remain wholly state forces. Like General Walsh, most other members of the Adjutants General Association did not like the attitude of the federal government in regard to state forces.[31] The friction with federal authorities could be quite strong at times. Brig. Gen. Mervin G. McConnel of Idaho related his personal experience with overzealous federal authorities. Several weeks before the meeting, an agent from the Federal Bureau of Investigation had come into his office at the state capital and challenged the legality of the uniform that General McConnel had chosen for the Idaho State Guard. In response, General McConnel showed the agent the *Army Regulation* on the subject and an example of the Idaho uniform to demonstrate that he had obeyed the letter and spirit of the law. Nevertheless, the experience had left him angry at the high-handedness of federal officials as they attempted to tell states what they could and could not do.[32] He wanted a clarification of the federal and state relationship in responsibilities and authority over the State Guard. The conference would not clarify that relationship, but it would allow the adjutants general to reach a consensus on what they wanted from the federal government, which was more material support but less interference.

The attending adjutants general had other areas of friction with the army. They debated about whether the federal government had any business inspecting a force defined as strictly a state institution. Here the adjutants general were on shaky ground. The receipt of federal property, especially arms, no matter how few or old, gave the federal government the right to inspect the conditions of storage and use of that property, and the Constitution specifically gave to the federal government the authority to organize the militia and prescribe discipline for it.[33] If the federal government had the constitutional authority to prescribe the discipline for the militia, then the federal government also had the right to inspect the militia to ensure that the states enforced that discipline. These points did not come up. The men of the Adjutants General Association were in no mood for legal hair-splitting. The federal government had taken their National Guard, then urged them to create new forces while the federal government gobbled up all available men and resources.[34] The recall of federal rifles was only the latest, although greatest, insult to date. And perhaps also, the federal government gave them a target for their frustration. The adjutants general, most of whom had served in the Great War and who had devoted much of their adult lives to the National Guard, remained at home while the units they had nurtured moved on to battlefields around the globe. Their new state forces, under-armed, under-equipped, and under-manned, some still in the planning stage, would be their only commands as the world entered a new world war even larger than the first. As a group they opposed further federal inroads into state forces, which would take away much of their remaining power and prestige.

What they wanted during the war was similar to what they and the National Guard Association had lobbied to get for years regarding the National Guard—federal support without federal control. The Adjutants General Association could be passionate in its defense of states' rights, as long as the federal government would supply money and equipment. State governments were too poor and fickle for the adjutants general to depend on them for a stable force. If adjutants general were to have uniformity among their troops and commands worth commanding, they needed federal weapons, equipment, and, most of all, money. What they passionately did not want was federal interference.

While the heads of state military departments attempted to present a united front to the War Department, other men in communities around the nation drilled and trained in local militia units that had little state and no federal support. For many men and states, the normal State Guard did not offer enough opportunities for service to the state and community. Sparsely populated rural areas in particular often lacked a State Guard organization. To fill this gap more than half the states allowed some forces organized on a more decentralized plan than the State Guard. In most states, these forces took the name of State Guard Reserve, although that term had other meanings in some states.[35] The states played only a minimal role in creating these forces. From the first days of American involvement in the war until the end, unknowable numbers of militia groups formed with and without official support.[36] Men were afraid and needed the comfort of organizations to dampen that fear.

The War Department soon took notice of these militia groups. General Williams, Chief of the National Guard Bureau, noted that in all regions of the nation, private military groups were forming, usually with local support and in one case with the encouragement of a corps area commander. In many states, local American Legion and VFW posts, as well as the National Rifle Association (NRA), encouraged their members to form such groups.[37] General Williams feared that the employment of such groups, and even their very existence, could lead to problems of authority and standing under international law regarding combatants. Problems could also arise between these groups and state officials. In one case, an argument between a member of a bona fide State Guard and a member of one of these other groups led to murder.[38] In January, 1942, during the informal discussions in the War Department on the desirability of federalizing the State Guard, the existence of these militia groups that were outside of the State Guard structure was brought up. During the discussions, the War Department hit upon the idea of creating what it termed a "Reserve Militia" to "avoid the manifest anomaly otherwise involved in having organized bodies in the so-called unorganized militia."[39] By the end of February, the War Department had outlined plans for such locally organized forces. Although the War Department believed that these militia groups might have some merit in channeling enthusiasm and patrolling rural areas, the impetus for creating and maintaining these forces remained on the local level. However, the War Department did suggest that if a state chose to recognize such forces they be given the name of "Local Defense Force," or "State Guard Reserve," which would differentiate them from the regular State Guard. Nevertheless, the War Department remained opposed to the idea of any militia over which a state adjutant general did not exercise command. For this reason, it advised that any militia group be incorporated into the State Guard structure. Even as a component of the State Guard, the War Department would not supply federal equipment or trainers for these forces.[40]

The War Department's main concern with these State Guard Reserve forces was that they be a state responsibility in their entirety, but the War Department desired states to exercise that responsibility. It did not want unaffiliated militia groups loose in the nation. Nor did state officials, who soon began incorporating these groups into their State Guard. In Maryland, questions arose on whether these militia groups would have the same standing at the War Department as did its regular State Guard. The state's Office of Civilian Defense sent to the National Guard Bureau a sample of the enlistment application it had been sending to potential members of the reserve militia, to see if retired officers could join. The NGB replied that reserve or retired officers of the army could join without jeopardizing their federal standing, as with the State Guard. The application asked the prospective militiaman, among other things, to list his membership in any rifle or gun clubs, and the number and type of firearms he owned.[41] As was the case with other states, Maryland expected members of the reserve to arm themselves.

Still, many states remained uncertain of the War Department's policy toward any State Guard units in excess of the strength that would receive federal support. In response, General Williams sent to all states a memo explaining recent War Department policies on the State Guard. Prominent among his points was that the War Department had no objections to the states keeping additional local forces for home defense provided they were "legally and officially affiliated with the authorized State Guards. These groups [were] not considered with allotted strength for assistance from the federal government."[42]

In effect, the War Department exercised a sort of negative leadership on the State Guard Reserves. It had no desire to forbid or discourage their existence, but it declined to incorporate them into any centralized planning, or to provide them any material support. Into this vacuum stepped an organization perhaps more suited than the War Department to assist small, quasi-independent militia units—the National Rifle Association.

In 1942, the NRA published its *Practical Home Guard Organization for Reserve Militia or "Minute Men"* to provide the general guidance and doctrine that the War Department had declined to give.[43] This short book outlined the NRA's plan for gun and hunting clubs to convert themselves into militia groups specifically to fight as a guerrilla force against any enemy that invaded the United States. The guide was based on the NRA's interpretation of federal law pertaining to militia, which said in part that a governor could organize the entire state as a militia if he wanted, only that the federal government limited the amount of equipment it could or would provide. The NRA's plan for "Minute Men" assumed the men would arm themselves, but strongly suggested that they be an adjunct to the State Guard. The NRA sought for such organizations men familiar with the countryside, who would be useful in preventing the formation of hostile groups or in detecting enemy penetration in sparsely populated areas. The NRA advised that men in these forces identify themselves at least by distinctive arm bands and caps, although uniforms were more desirable.

The NRA desired sporting and hunting clubs to enroll in these Minute Men groups as a unit and form part of the state's militia under the control of the adjutant general of each state rather than be placed under the office of civilian defense. This was part of the NRA's stress on the force as militia and not simply a group of overage Boy Scouts who wanted to help. Minute Men were to be semi-guerrilla in nature. The NRA stressed that rather than drill the men should work on marksmanship and the ability to cooperate with each other. As a follow-up, the NRA published *Practical Home Guard Organization, Supplementary Bulletin Covering Application of the "Minute Man" to Large Cities,* which further adapted the original "Minute Man" plan to urban areas.[44] Unless authorities saw sabotage as imminent, Minute Men were not be used as watchmen, nor should they be used for strike duty, but were to hold themselves available for use as posses.[45] For the Minute Men, the NRA sought the more romantic and exciting roles of militia and sought to distance the Minute Men from the more controversial or dull duties. They saw these Minute Men, or State Guard Reserve as they were officially called in most states, as a force for purely local defense, with the men in it knowing all the people and land in their area. With these groups organized throughout the nation, the NRA believed the nation to be unassailable.

But the NRA did not champion the idea that every American had an automatic right to carry weapons as part of the militia. The pamphlet specifically stated that

> NO OFFICER OR MAN WILL BE PERMITTED TO CARRY A GUN ON MILITIA DUTY UNTIL HE HAS SATISFIED THE COUNTY COMMANDER, OR COUNTY SMALL ARMS INSTRUCTOR THAT HE IS QUALIFIED TO HANDLE HIS WEAPON SAFELY AND EFFICIENTLY![46]

The NRA stressed that every county government had the right and the duty to ensure each man's competency with weapons before permitting him to bear arms as part of the militia.

No state ever adopted the NRA plans outright, but many allowed forces organized along similar lines. Some, in fact, adopted the name "Minute Men" for their State Guard Reserve, but a variety of names came to describe these local militia forces. In general, such companies depended on the personality and imagination of their company commander, who usually recruited his own company. In most states, a monthly status report was the sum of contact between the company commander of a Reserve unit and state military authorities. Typical of such forces was Virginia's Reserve Militia, organized in May, 1942. Sportsmen, skeet shooters, the Isaac Walton League, and other similar groups formed their own companies within the Reserve Militia, which could not be sent outside of their home county. By the end of 1942, the Reserve Militia of Virginia counted about 5,000 members, including three cavalry troops whose members supplied their own horses. During its existence, members of the Virginia Reserve Militia assisted the Federal Bureau of Investigation by reporting on suspected subversive activities and locating deserters from the federal military.[47] Reserve Militiamen also looked for a missing trapper, who was later found dead from a gunshot wound, helped a sheriff arrest two men, protected a wrecked army plane, searched for missing children, and provided security for a train car filled with government supplies. They searched for a suicidal veteran who had been discharged for mental trouble, whom they found after he had shot himself in the head and died despite their efforts.[48] When the governor disbanded the Reserve Militia on September 15, 1945, many of its members transferred to the State Guard.[49] But that lay far in the future. In the meantime, state military authorities groped their way toward support and a role for state forces.

In this endeavor reserve units of the State Guard were almost forgotten, as state military leaders focused on preparing their regular State Guards for additional burdens.

The joint meeting of the Adjutants General Association and the National Guard Association at Harrisburg, Pennsylvania, on April 1–3, 1943, was the first meeting of the National Guard Association since October, 1940. The Executive Council of the National Guard Association had continued to meet annually in the intervening years, but in reality, the goals of the two associations were so interwoven—and the adjutants general as a rule were all members of the National Guard Association—that the annual meetings of the adjutants general had served as rump sessions of the National Guard Association. The joint associations recognized, as did the War Department, that the numbers of men and units in the State Guard were small at the present, but as more federal units left the United States for deployment overseas, the need for the State Guard would increase and take on added importance.[50] The assembled proponents of state soldiery recognized that the presence of large numbers of federal troops scattered around the nation in various training camps made state forces somewhat superfluous in the early years of the war. In the event of disorder, almost every governor could call for the service of the federal soldiers in his state, whether the War Department wanted this or not. But the members of the associations also recognized that the vast majority of federal soldiers would soon be leaving for service overseas, and the nation would need to rely more heavily on state forces in the event of domestic crises.

Recognizing this future reliance on the State Guard, the associations decided that, although they opposed any plan that would place the State Guard under direct federal control, they did desire lavish federal support for the State Guard.[51] One of the resolutions passed at the conference advised the federal government that support for the State Guard should not exceed that given to the National Guard before its induction but also reiterated the right of states to create as many units as they wanted.[52] This stand was disingenuous, for the federal government had not been demanding that states create a force larger than the prewar National Guard, and in fact had decided to provide arms to the states for a force half the strength of the pre-induction National Guard. The War Department's policy made sense. The National Guard's primary mission was to fight wars, with state missions as a secondary role, whereas the State

Guard existed only to serve the states. Despite the complete lack of a federal mission for the State Guard, the associations still lobbied the federal government to provide equipment, uniforms, rifles, bayonets, and slings for each enlisted man, sidearms for officers, and training funds for 180,000 State Guardsmen.[53] The funds were to pay for training equipment, transportation, and food rather than to pay the individual State Guardsmen.

The second morning of the convention, Friday, featured an address by Col. W. F. Adama, the executive officer of the National Guard Bureau. He quickly went to the crux of the problem with the State Guard. The many different states and territories, the federal government, and even individual State Guardsmen, often had conflicting concepts of the role and responsibility of the State Guard. Some saw its mission as one of maintaining civil order, being the state governments' final recourse when situations exceeded the abilities of ordinary police forces. Others saw its mission as the patrolling of coasts, bridges, and other vital points to thwart saboteurs. Those wanting a more heroic role saw the State Guard fighting a guerrilla war behind enemy lines in case of invasion.[54] The War Department remained ambiguous about missions for the State Guard. Colonel Adama explained that the War Department preferred to let individual states create the type of forces they wanted and only wanted states to let it know what they were doing.[55] For the first year of the existence of the State Guard, from when the National Guard entered federal service until the entry of the nation into the war, the War Department had not taken much notice of the State Guard.

Until Pearl Harbor, states had gotten little military assistance from the War Department, which, according to Colonel Adama, had little interest in the State Guard until then. States that had put their State Guards on active service following Pearl Harbor had received the bulk of federal equipment issued early in the war.[56] A large percentage of the early federal support went to California, which was particularly fearful of attack and sabotage. However, this federal support proved to be short-lived, and regulations soon killed it. After the Quartermaster Department of the army had issued $2,000,000 worth of obsolete Lend-Lease and Civilian Conservation Corps equipment, as well as another $2,800,000 worth of Regular Army equipment, the Budget Bureau, backed by the judge advocate general, informed the quartermaster general that the issue of the supplies had been illegal, and because of this, the quartermaster would not be able to recoup in the next budget the money it had already spent on state forces. Of course, this ruling temporarily ended the issuance of federal equipment to state forces.[57] Not that the few uniforms the federal government had managed to issue before that source was cut off made much difference. Most State Guardsmen, especially in the early years, were too old for military service, men in middle age. But most of the clothing that they received from the War Department came from former CCC stocks, which had been created to clothe young men and teenagers, with the more common sizes being worn out during the existence of the CCC. As a result, most of the uniforms were much too small for State Guardsmen or, in some cases, too large.[58]

Still, the National Guard Bureau remained busy throughout the war trying to equip the State Guard. An amendment to the *National Defense Act* on October 1, 1942, allowed the secretary of war to issue some items to the State Guard, but legal experts in the War Department ruled that this still could not be done because no one was authorized to appropriate the money for the equipment for the Secretary of War to issue. The point was moot anyway. State Guard units had no official Table of Organization, so they also lacked a Table of Basic Allowances, meaning that no supply program could be developed within the War Department bureaucracy.

In response to these bureaucratic obstacles, the National Guard Bureau developed a flexible Table of Organization for the State Guard that could be grafted onto most State Guard organizations already in existence. It called for a battalion with as many companies as a state desired. It also created plans for a so-called Tactical Headquarters, which functioned like a regimental headquarters, but lacked the service and supply elements.[59] Several states adopted some of these provisions. The Bureau also developed a State Guard manual, but the Division of Services of Supply refused to publish it on the grounds that it had taken twenty years to get rid of an ROTC manual that it had published many years earlier and did not want to be saddled with piles of State Guard manuals. The *Infantry Journal* expressed some interest in publishing it for sale,[60] although the manual eventually was published by the Military Service Publishing Company as *The State Defense Force Manual*.[61] These petty obstacles for the State Guard underscored its lack of standing within the War Department.

Despite the obvious links between the State Guard and the National Guard Bureau, the two did not always have a clear relationship to each other. Just before Pearl Harbor, an army staff study recommended taking all but record keeping functions of the State Guard away from the National Guard Bureau and giving it to the army's adjutant general and provost marshal general, on the grounds that they had the most interest in the State Guard because of their inherent internal security orientation.[62] The Adjutants General Association and the National Guard Association disagreed with the staff study, and recommended that all matters of the State Guard be supervised and coordinated by the National Guard Bureau, which made sense because the National Guard Bureau had succeeded the former Militia Bureau.[63] The *National Defense Act of 1916* had attempted to make the terms "National Guard" and "organized militia" synonymous, but the induction of the National Guard had changed that. Part-time state military forces were by definition militia, and placing them under the section of the War Department originally created to deal with militia matters seemed the most logical place for them.

Placing the State Guard under the National Guard Bureau would also give the National Guard Association a means to influence War Department policy relating to it. Since the creation of the National Guard Bureau, it had worked hand in glove with the National Guard Association, being in effect the vehicle through which the National Guard Association worked with the War Department on issues relating to the National Guard, without the need to use its lobbying power in Congress.[64] To continue this cozy relationship would allow the National Guard Association to exercise its traditional influence on militia matters during the war while preventing the State Guard from becoming a threat to the standing of the National Guard after the war. It would also give a mission to the National Guard Bureau, which, since the induction of the National Guard, had little else to do.

Finding a reason to exist would prove difficult for the National Guard Bureau during the war. The reorganization of the War Department in March, 1942, divided the army into Ground Forces, Air Forces, and Service Forces. The National Guard Bureau, greatly diminished in importance, had been downgraded from a special staff, and came instead under the direction of the army's adjutant general, who then delegated supervision of the State Guard to the National Guard Bureau. But the war had robbed the NGB of most of its power, and it would be bounced around the War Department until it and the National Guard regained their original position after the war.[65] Most of the reshuffling had negligible effect on the State Guards, for they remained a state force, and thus in reality fell under their state adjutants general, who answered to their governor. The War Department's role was one of providing guidelines and policies—suggestions, really—and in tracking the progress of State Guard training and equipping. Guidelines and policies could be provided cheaply; equipment proved much more dear.

In midsummer 1942, the federal government assigned to the State Guard a priority for obtaining goods that placed them next below the lowest ratings for the federal forces.[66] Every unit of the U.S. Army, no matter how far removed from the war effort, had a higher priority to receive material from the War Department or place demands on the economy than the busiest State Guard unit. With the expanding federal forces running short of supplies, this rating seriously hampered any chances of state forces receiving equipment until relatively late in the war. The deliberations of the Adjutants General Association and the National Guard Association were filled with more issues than just equipping the State Guard and defining its role vis-à-vis the War Department, but the State Guard remained their most immediate issue.

The hothouse atmosphere of the war years required the adjutants general to meet in Ohio two months after the Harrisburg meeting to resolve other problems. One of the points for discussion concerned a matter that had been the bane of state forces in both world wars, and indeed ever since—the distinctness of state military uniforms. At the time, the National Guard Bureau was floating an idea for a distinctive badge to be worn by members of the State Guard.[67] The issue could be solved by the language used. State Guardsmen often expressed their desire for some uniform device that indicated long or emergency service in state units, but they loathed federally directed markings that the state soldiers interpreted as indications of a second-class status. During the Great War, the War Department had required state forces to wear a red felt star on their sleeves, which became much hated objects of derision. State military leaders wanted a uniform that was available during the wartime economy but that state soldiers would wear proudly. The War Department wanted the states to adopt a uniform that would not soak up needed resources and would be unmistakably different from any federal military uniforms. The issue could not be solved in a meeting room in Ohio, but the state soldiers had staked out a unified position for the War Department.[68]

Protecting the uniqueness of the army uniform concerned the army greatly in the early days of the war. In January, 1941, Secretary of War

Henry L. Stimson rescinded the authority of a private group called the Military Order of the Guards, Incorporated, to wear its uniform, which was similar to the uniform of the U.S. Army. The War Department's main interest in the Order stemmed from the "promiscuous" wearing of the army uniform by "unauthorized groups."[69] With established private groups losing authority to wear their uniforms, the uniforms of state military forces came under increasing scrutiny.

As the War Department sought to protect federal uniforms from infringement by private organizations, it also strove to maintain the distinctions between the uniforms of state and federal forces. The issue of protection of the uniform had received only minor interest in the interwar period, but with the beginning of World War II, and the buildup of the U.S. military, the War Department took an active role in maintaining the distinctness of federal forces. In January, 1941, the NGB decided that the red cloth stars would no longer be required on uniforms of state forces.[70] In reality, this edict affected few states, such as Maryland, which required their wear. The regulation required their attachment only to uniforms that were similar to a federal uniform, and all states claimed their uniform to be distinct. The War Department authorized, but did not require, a distinctive scarlet sleeve braid for State Guard uniforms. Whether used or not, the uniform of state forces had to be recognizably different from the uniform worn by members of the U.S. Army, Navy, or Marines. Still, states could not compete with the federal government in uniform procurement. The adjutant general of Virginia requested that a higher priority be granted for two hundred gold-plated collar insignia for his State Guard officers.[71] Even this was denied and he was told to use a noncritical material.[72] The fatigue uniform worn by most enlisted men presented less of a problem. States that did not design a uniform specifically for their forces usually used obsolete federal stocks. These could be the Great War style uniforms with leggings and the M1917 steel helmet, or, more often, former CCC uniforms. State Guardsmen patrolling the local dam in the spruce green uniforms of the CCC with plastic badges were readily distinct from federal forces.

For state officers, the uniform procurement remained an issue throughout the war. In some states, only officers wore dress uniforms. The uniform states adopted for their officers was usually similar to that worn by members in the army, but with a few alterations. For officers who had only recently left the National Guard, they could simply make a few changes to their National Guard uniforms and be properly outfitted. For others, simply getting a dress uniform could be a minor headache. The federal forces had priority over all other concerns, meaning in practical terms that individual State Guard officers often could not get uniforms through private suppliers. Instead, they tried to use the Military Clothing Sales stores located on all major army installations. After an initial period of uncertainty by the officials who ran the Military Clothing Stores, the War Department ruled that State Guard officers could purchase their dress uniforms there, but the seller needed to remove all buttons and other items that identified a federal uniform.[73]

The War Department had more interest in State Guard uniforms beyond their appearance and acquisition; it also attempted to control the wearing of State Guard uniforms outside of a member's home state. State Guardsmen sought War Department permission to wear their uniforms beyond the borders of the United States. After the British government invited some members of the Maryland State Guard to go to England to study the British Home Guard in 1943, the Marylanders requested permission to go in their state uniforms. In 1944, the Adjutant General of Texas wanted to send a battalion from Texas to participate in the Military Parade in Reynosa Tampa, Mexico. The adjutant general of New York wanted to send the Sixth Regiment into Canada, without their arms, to participate in the Dominion Day Parade.[74] For bona fide military events, the War Department usually acquiesced. What it did not want was state military personnel wearing their uniforms for no other reason than to impress the locals.

After the uniform issue, the adjutants general moved on to a subject that had agitated the National Guard in the late nineteenth and early twentieth centuries—federal pay. In line with their constant desire for federal support without federal control, members of the association raised the issue of federal subsidies for State Guards on active duty, either for training or service.[75] But on this issue, the adjutants general found consensus evasive. Federal funds would have been nice, but the struggle for federal pay for the National Guard had lasted decades. No members of the Adjutant Generals Association or National Guard Association desired the State Guard to exist much beyond the return of peace, lest it become a threat to the reconstitution of the National Guard. Federal pay for strictly state forces could entrench their role in the postwar period, and so most Adjutants General shied away from requesting such pay.[76]

More surprising was their reaction to a proposed federal plan to credit service in the State Guard for pay purposes in the federal military. This plan would give men in the federal forces longevity pay based on time spent in state forces as well as in federal forces. The association decided to oppose this plan on the grounds that it would overload state staffs already swamped with work. In those days, before a centralized military pay system, state staffs spent an inordinate number of hours validating longevity pay for National Guardsmen after they entered federal service. The adjutants general had no desire to increase this burden by adding State Guardsmen to their lists. In spite of this ruling, Brig. Gen. Vivian B. Collins of Florida claimed that federal finance officers in his state were already crediting State Guard service for longevity pay. However, Collins may have been mistaken, or perhaps the federal soldiers who computed pay for Floridians in federal service had misunderstood the differences in the State and National Guard, and federal soldiers who had earlier served in the Florida State Guard got unexpected bonuses with their monthly federal pay.[77]

Other than these issues, the extra meeting for 1943 covered little of importance. They spent part of the remaining time discussing Army Regulation 850-250, *Regulations for State Guard,* but reached no new conclusions.[78] However, the State Guard was about to leave its infancy. Much of the vast federal military had already left the United States, and much of what remained would soon follow. The State Guard was about to receive increased responsibilities for the home front, even if the threat of invasion grew less likely with every passing month. With these new responsibilities came increased federal support, until the State Guard began to look more like the prewar National Guard. That said, the State Guard remained more dependent on local and state direction, and so the State Guard of each region remained distinct in its size, outlook, and character.

CHAPTER 7
The State Guard Readies for Action

In the lobby of the modern headquarters of the National Guard Association in Washington, D.C., stands a display case containing a six-inch-tall figurine from each of fifty-three states and territories, dressed in a costume representing an early militia force of the state or territory. Alaska's figurine is of an Eskimo in the Alaska Territorial Guard. It is the only representative from a force raised to replace the National Guard. Although the Alaska National Guard had come into existence before World War II, and the territory had a home guard in the Great War, the Territorial Guard of World War II captured the public's imagination.

Alaska had thousands of miles of trackless territory that concerned federal defense planners in the years between the world wars as they considered the possibility of a war with Japan. The immense territory could absorb endless numbers of soldiers if the army were to patrol it. The Alaska National Guard, a new force at the beginning of World War II, contained only four companies, and it entered federal service with the rest of the National Guard.[1] To replace it, Alaska began forming its Territorial Guard to protect the trackless expanse. For many Alaskans, the Territorial Guard would be their first experience with militia. Creation of the Alaska Territorial Guard began shortly after Pearl Harbor. Originally, territorial leaders envisioned a paid militia force of eight hundred men geared toward controlling civil disorder in the cities. During the Great War, fearing industrial sabotage, the territory had created a home guard in many of the larger towns in order to cow foreign workers whose homelands were then at war with the United States. However, the dangers faced by Alaska in World War II were markedly different from those in the Great War. In 1941,

Alaska became acutely aware of the vulnerability of its coastlines to invasion by the Japanese. The immense size of Alaska made the planned eight-hundred-man force almost useless, and the army was not keen on sending north the thousands of soldiers that a proper defense would require. In response, Gov. Ernest Gruening planned for all adult males who were not in the National Guard, the federal military, or performing vital work, to be enrolled in the Alaska Territorial Guard. In an example of deft leadership, the governor was able to create a force of both Native Alaskans (Indians and Eskimos) as well as white Alaskans, two groups that did not always work together harmoniously.

The governor made his first trips ever into the interior in order to ask the Eskimos to enlist. They enlisted almost to a man. In return, to identify the members of the Alaska Territorial Guard, the governor gave them each a shoulder patch that cost fourteen cents. Around twelve thousand patches were eventually issued. Upon joining the Territorial Guard, the men were told to report any sightings of strange men, boats, or planes. But they would be more than passive scouts. They were instructed to shoot any Japanese who tried to land by boat or plane. In the event of invasion, the Natives were expected to act as a guerrilla force, becoming the eyes and ears of the army and constantly harassing the enemy. Concerning the Territorial Guard's plans for guerrilla warfare, the National Guard Bureau noted that while the War Department was not assigning combat missions to State Guards, it had no problem if commanders or state officials wanted to assign such missions to their forces.[2] The governor told the Natives that the Japanese had bombed Pearl Harbor and had also dropped bombs on Alaska, a reference to the two Japanese air raids on the American base at Dutch Harbor, on the island of Unalaska, near the eastern end of the Aleutian chain, which occurred as a corollary to the Midway campaign in late spring 1942.[3] Although he also warned them that more Japanese would come, he most likely remained unaware that the unpopulated islands of Kiska and Attu, on the western extreme of the Aleutians, were at that moment occupied by Japanese soldiers. The governor tried to bring the danger closer to the core of Eskimo culture and livelihood, by telling them that the Japanese would come to take the fish, whales, and seals, and leave none for the Eskimos.

The Eskimos who joined the Territorial Guard became the legendary Eskimo Scouts, which was mostly the creation of Maj. Marvin "Muktuk" Marvin, an Air Corps officer assigned by the War Department to assist the governor. Marvin had the responsibility for the northern parts of Alaska and became popular with the Natives.[4] Some whites did not like the idea of armed and organized Native Alaskans, but Marvin opened the eyes of the Natives to their American citizenship and became an advocate for their rights.[5]

Alaska was not alone in its apprehension over the vulnerability of its coasts to a Japanese attack. All states and territories bordering the Pacific shared this fear, and thus the creation, training, and employment of State and Territorial Guards took on an urgency not matched on the east or Gulf coasts, or in the interior. When the five thousand officers and men of the Washington National Guard entered federal service in January, 1941, a State Guard of equal number replaced them, as required by state law. This new force, the Washington State Guard, was formed purely to assume the state missions of the National Guard. However, by early 1941, the adjutant general, Maj. Gen. Walter J. Delong, decided that the probability of the entry of the United States into the war, which included the possibility of invasion, meant that the State Guard needed to be reorganized with a posture more suited to opposing Japanese attacks. He temporarily suspended recruiting, discarded the original plans, and transferred the personnel to what he termed the "State Guard Reserve," which was a list of State Guardsmen rather than units. There they were to be used as a nucleus of a new force.

Organization of this new force began on June 17, 1941, with the creation of the Fourth Washington Volunteer Infantry Regiment. To arm the regiment, the state asked for and received 2,202 Enfield rifles on loan from the federal government but had to pay for the ammunition. The state provided each man in the State Guard an olive-green one-piece coverall uniform with a distinctive insignia for the cap and sleeve, and a heavier woolen uniform with leggings for inclement weather and field duty. The regiment was completed on paper by December, 1941, when each company was authorized to recruit a third platoon, an expansion that gave each rifle company four officers and 176 enlisted men. Even with the expansion, the regiment still had fewer units than the adjutant general desired, and it left many armories empty. To fill the vacuum, and to bring the State Guard up to its allotted strength, the adjutant general organized the First Provisional Infantry Battalion with 10 rifle companies of 3 officers and 120 enlisted men each. Early in 1942, the Washington State Guard increased its numbers to 4,017 enlisted men, the strength of the prewar

discharged men petitioned the governor for the right to serve. These men averaged twenty years of age, and some of them had college degrees, but most of them were still university students when the war began. Accepted as a labor battalion under civil service, they were attached to the Thirty-fourth Combat Engineers Regiment, where they performed heavy manual labor. Because of the high percentage of college men in the unit, they called themselves the VVVs, which stood for "Varsity Victory Volunteers," although their official name was the Corps of Engineers Auxiliary. For eleven months the regiment stayed at Schofield Barracks, on Oahu, where they quarried rock, strung barbed wire, and built military installations, roads, warehouses, and dumps. Eventually, they requested and received the deactivation of the regiment in order to join the army, with most of them ending up in the legendary 442nd Regimental Combat Team.[26]

Despite the attitude a few years earlier that the presence of so many regular military forces on the islands made a National Guard unnecessary, the attack had made the islanders feel very vulnerable. As a result of the geography of the territory—many islands scattered over a vast expanse of sea far removed from the mainland—Hawaii created a more extensive militia system than any other state or territory. Unlike the rest of the nation, where a State Guard and maybe a State Guard Reserve comprised all the opportunities for militia service, Hawaii's system was much more comprehensive. One element of the Territorial Guard was the Organized Defense Volunteers. This force had a standing unique to American forces in World War II. Although they were essentially civilians, as a unit they formed part of the armed forces of the Central Pacific area. In function and standing they resembled more the British Home Guards rather than the American State Guard.[27] Members took an oath to serve on their home island whenever the commanding general of the Central Pacific believed an invasion was imminent. They were also liable for induction as a unit or as individuals into the U.S. Army. Their numbers peaked in 1942, when some twenty thousand men belonged to the force. Missions included beach defense, watching strategic and vulnerable points such as hilltops, runways, and crossroads, traffic control, providing guides and scouts for the army, and, if all else failed, implementation of scorched earth in the path of invaders. They were last-ditch soldiers. In preparation for their role in an invasion, they drilled every Sunday morning under the direction of soldiers detailed from the army. In addition to training, they also took turns at sentry duty every night. On the larger islands of Maui, Molokai, Kauai, and Hawaii, the Organized Defense Volunteers were backed by rural "mounties" in ranch country, who assisted them on horseback. The name of the Defense Volunteers changed throughout the war, as did its training and missions. Still, interest in maintaining the force allowed its continuation long after the threat of invasion had passed. Nine units remained in existence all the way until the formal inactivation on July 4, 1945. Of the nine, five units were manned by rural plantation workers.

All of the larger islands had some form of militia. On the "big island," the Hawaii Rifles were organized in February, 1942, absorbed the Provisional Police of Hilo, and contained at peak strength 4,500 men. During exercises held by the army in July, 1942, to test the ability of the Rifles to respond to an invasion, the First Battalion completed a night march of twenty-nine miles without losing a man over terrain considered impassible.[28] The Kauai Volunteers formed in March, 1942, with 2,400 members, 90 percent of whom were Filipinos. Another 3,000 men joined the Maui Volunteers and its offshoot, the Molokai-Lanai Volunteers. One-third of the adult male population of Lanai belonged to it. The battalion at Lahaina, on Maui, included a labor auxiliary of men of Japanese descent, who were not allowed to join the Volunteers but who were organized to provide emergency repairs, demolition, clearance, and to fight fires in the cane fields. In January, 1944, the Hawaii Scouts and Oahu Scouts were combined to become the Oahu Volunteer Infantry.[29]

Honolulu, the largest city in the territory, had three militia groups of its own. The Businessmen's Military Training Corps contained white men and men of mixed white and Native Hawaiian ancestry. Within two weeks of its formation in January, 1942, this force contained 1,500 men organized into seventeen companies. The Businessmen's Military Training Corps attracted many prominent men, whose average age was forty-two. This group had the oldest average age of all the militia groups in the territory. One private was a seventy-six-year-old former chief justice of the territory. Two-thirds of the men in it were Great War veterans. Their mission was to immobilize enemy aliens in Honolulu in case of invasion. The group gathered data on aliens and maintained a ten-foot-high map showing vital places in the city, as well as areas of where residents of Japanese descent were concentrated. They were as much concerned with what they saw as a potential enemy from within as the real enemy from without. The exclusionary nature of the Businessmen's Military Training Corps led to demands from nonwhites for permission to form their own militia. From this desire came the Hawaii Defense Volunteers. At its first training meeting on May 24, 1942, some 130 men attended, which grew to include 800 men by November, 1943. Most of the men in it were Chinese, but it also included Filipinos, Hawaiians, Puerto Ricans, Koreans, and even a few whites.[30]

In August, 1942, women employees of the Office of the Military Governor created the Women's Army Volunteer Corps, which included about 400 women who were civilian employees in the various army offices. They formed to bear some of the local security burden in order to release men for active service in an emergency. The Women's Army Volunteer Corps received federal recognition as a noncombatant component of the Organized Defense Reserves after the organization of the Women's Army Corps (WACS). Each member undertook some basic training, participated in reviews, and held two-day bivouacs. Their actual duties were less exciting if still necessary—they performed emergency stenographic work, drove army buses, ushered at army performances, visited the wounded at hospitals, and assisted the tax office and war agencies.

The Territorial Guard, the various militia on each island, and the Women's Corps did not exhaust all the opportunities to serve on the islands. In October, 1942, 1,500 men formed the Hawaii Air Depot Volunteer Corps to help around Hickam Field with chemical decontamination, fire fighting, first aid, antisabotage, evacuation of noncombatants, and sentry duty.[31] This was probably the only volunteer militia in the nation to get antiaircraft machine guns in World War II.

Because the territorial government did not have nearly enough arms for all these forces, most members armed themselves. Many of the Filipinos used bolo knives, which they fashioned from car leaf springs. Only later in the war did federal arms arrive, including .45 caliber pistols, Enfield rifles, tommy guns, and later, shotguns. The American Legion and other organizations, plantations, and businesses helped to pay for uniforms for the various militia groups, but often militia men paid for telephone service, stationery, and other operating supplies with their own money. Not until 1943 did the territorial legislature provide insurance for members injured or killed on duty. Before that the men served completely at their own risk.

The army estimated the strength of the various Hawaiian militia groups at about a division. Lt. Gen. Robert C. Richardson, Jr., the commanding general of Army Forces, Central Pacific, had enough faith in their ability to protect Hawaii that he redirected many of the army soldiers that had originally been slated for garrison duty on the territory instead to take part in the attack on Makin Island, in the Gilberts, in November, 1943. Other army troops were withdrawn for offensive actions elsewhere in the Pacific Theater. From October, 1943, through July, 1944, the Businessmen's Military Training Corps and the Hawaii Defense Volunteers relieved army sentries from duty around Honolulu every evening between 1800 and 2400 hours. The Hawaii Air Depot Volunteer Corps spent 16,000 hours protecting air depot installations beginning on December 1, 1943, and remaining into 1945.[32]

Despite the assumption of responsibilities, interest among the Volunteers waned after the first months and the lessening of the threat of invasion. This was especially pronounced on Oahu, the island with Pearl Harbor, Hickam Field, and most of the federal military presence in the territory. Business responsibilities, labor shortages, and induction into federal forces took their toll. Physical examinations in 1943 found 40 percent of the militia members were disqualified for service. The army later ordered the discharge of all men above the age of sixty. In May, 1944, and January, 1945, the army considered disbanding all units but found that the groups provided needed services in the islands.[33]

The West Coast experience of large State Guard forces poised to oppose an invasion was not the only model of the State Guard. Small, lightly populated states, or states with few or no urban industrial areas usually created much smaller State Guard forces with more limited missions. Typical of these states was Delaware. Although situated along the Atlantic coast, the state created a small State Guard with little expectation that it might one day face a seaborne invasion. Instead, Delaware entrusted the defense of its 110 miles of coast to the federal forces. The federal government had long recognized the vulnerability of the Delaware coast and in late March, 1941, began construction of what would become Fort Miles on Cape Henlopen, at the mouth of Delaware Bay.[34] Nevertheless, the state government recognized the need for an organized militia after the National Guard entered federal service. As a result, the Delaware General Assembly passed the enabling legislation for the State Guard on April 14, 1941. By that time, approximately 2,500 men from the state were in federal service, with two-thirds in the National Guard.[35]

The Delaware State Guard of World War II was considerably larger than its Great War predecessor, although still small by national standards. It originally functioned as a single battalion but later expanded to a regiment of two small battalions, containing a total of between 450 and 500 men.[36] The men enlisted for a year, with reenlistment possible. The organization had seven line companies, plus a headquarters company, medical detachment, and a "Grenadier Platoon," trained in grenades, automatic weapons, and gas and smoke employment. In addition, the State Guard contained a band, under the direction of Captain J. Norris Robinson, who had been in National Guard bands since 1902.

Despite the presence of a band and a grenadier platoon seemingly geared toward public order, the State Guard saw itself as a combat unit and not as a police force. At its first encampment, in the summer of 1943, the State Guard practiced modern warfare techniques. The commanding officer of nearby Fort DuPont, Colonel George Ruhlen, visited for an inspection. Later, the state staff formulated plans for each unit to protect the vital installations—power plants, utilities, communications—near their home stations. However, the War Department reminded the state that it foresaw the only use of the State Guard in a combat role as responding to small raids, and saw the State Guard's mission mainly as internal security. By the summer encampment in 1945, the State Guard training emphasized more riot control training, including the use of gas.[37]

In many ways, the summer training encampments of the State Guard copied those of the National Guard, although lasting one week instead of two. As with the National Guard, the annual Governor's Day at camp became a highlight of the ceremonial side of the force.[38] Traditionally, a state governor would spend one day each year visiting his state's militia while it was in camp. The increases after World War II in size of the National Guard would later make this practice impractical, but the State Guard remained small enough for the visits to continue during the war. Additionally, as with the National Guard through the 1970s, the State Guard of World War II placed great emphasis on marksmanship training and, especially, competition.[39] Rifle matches, with medals awarded to winners, were a popular feature of the encampments, and gave the governor some pleasant activities to perform during his visits.

South Dakota, over thirty times the size of Delaware and situated far inland, had seemingly different security needs than small coastal Delaware, but it too created a small State Guard during World War II. In South Dakota, state law followed the pattern of most states: militia was classed as either National Guard or unorganized. However, when the United States entered the war, the government of South Dakota saw no need to change state law when their National Guard became part of the Army of the United States. Instead, South Dakota simply organized new companies in the early spring of 1942, which neither sought nor received federal recognition as National Guard.[40]

In all, the adjutant general organized a total of four companies in the towns of Aberdeen, Watertown, Pierre, and Sioux Falls. The capital, Pierre, in the geographic center of the state, was the westernmost town to get a company.[41] The other three were in the largest towns in the more heavily populated eastern half of the state. Even in South Dakota, militia was not a rural institution. Only the larger towns could support the required numbers of able-bodied men who could get to their armory for training. Each company had between 57 and 76 men. As with most states, former National Guard officers and Great War veterans filled the officer ranks of the South Dakota State Guard. Two of the initial four company commanders had been serving in the South Dakota National Guard at the time it had been inducted for a year of training but had later resigned, one for physical disability, the other for unstated reasons.[42] Unlike most states, South Dakota depended wholly on federal stocks to uniform and arm their force.[43] As a hedge against possible state use of the force, the legislature budgeted $25,000 for the force to use if needed.[44]

The western side of the state, in particular the Black Hills region in the southwest, was not ignored. Fire had long been the greatest enemy this region faced, and the state created a militia-type force to respond to this threat in the spring of 1942. Two line companies and one headquarters company completed this force, known as the "Fire Protection Force," or, as it was more commonly known, the "FPF." The federal government bore responsibility for most of the region, which included Custer National Forest. Partially for this reason, the federal government provided the equipment for the FPF, which assumed the role of nucleus of a larger force in the event of major fires in the Black Hills.[45] With the State Guard and the FPF, the adjutant general believed he could meet the needs of the state. The companies were well trained and ready for state service.

South Dakota had been steadily losing population for several years.

The economic boom of the early war years did not reverse the trend. Indeed, as an agricultural state, the war further depopulated the state. Men not in federal service tended to leave for areas of the country where jobs were available in war industries, of which the state had none. As a result, the South Dakota State Guard of World War II was much smaller that the home guard the state raised during the Great War, which had included over seven thousand men in forty-five companies.[46] In this war, the adjutant general stressed the lack of need for a larger force in an agricultural state and also recommended that no additional companies be formed, more from lack of response than from a well-crafted plan. He complained that "numerous cities have been offered State Guard units, but they don't seem interested."[47] South Dakota did not experience the large grassroots movement to "do something" that often created movements for some type of state force before state governments began to form them. South Dakota had no large urban concentrations of workers who might riot. Another factor in the adjutant general's confidence came from the federal troops within the state, which he assumed would be available in a crisis.[48] He seemed oblivious that his reliance on federal troops was the very situation the War Department sought to avoid.

Although some states had small State Guard forces, and some had more militia units than the federal government would support, four states, some of the territories, and the District of Columbia created no comparable militia forces to replace their National Guard. Arizona claimed to have neighborhood groups organized, which on signal would report to neighborhood assembly stations and move as a group to prearranged central posts. The state planned to organize them into squads, platoons, companies, and battalions. The force would not be drilling or parading under arms. Members were to be drawn mostly from veterans groups.[49] However, no evidence indicates that this plan ever reached implementation stage. Likewise, the governor of Oklahoma believed that four or five companies of infantry located strategically at armories should take care of any problems.[50] In 1941, a much larger force of three battalions reached the planning stage, with almost three thousand federal rifles requisitioned. However, no force was actually created because the state legislature failed to appropriate any money for the force, and the state instead simply increased its Highway Patrol.[51]

Nevada, which also never created a State Guard, had responded to an inquiry before the war from the National Guard Bureau by saying that the governor had the power to use state police to maintain peace and order until Congress authorized state military forces.[52] But after Congress did pass the necessary legislation, the state opted to forgo creating any force. In his *Annual Report* for the two years astride the entry of the nation into World War II, the adjutant general of Nevada, which had one of the newest and smallest National Guards, made no mention at all of the possibility of organizing a new militia. Instead, he reported mostly on his difficulties in running Selective Service, his quarrels with the United States Property and Disbursing Officer, and his shortage of office space. The creation of a State Guard does not seem to have occurred to him.[53] By the summer of 1944, little had changed in the state's military department, although the work level had decreased. The activities of his office had been largely curtailed by the induction of National Guard and he became the only adjutant general to claim a decrease in activity on account of the war and that his main business had been furnishing certificates of service to former National Guardsmen.[54]

Despite his lack of interest in a State Guard, as the war years passed the adjutant general became very upset by the implications inherent in plans for a large postwar standing army he believed the United States would create to take on the role of world policeman. This large standing army would be the death of the National Guard because the federal government would not be able to support both. He filled his reports with quotations from the Constitution to make his point that the nation must have a National Guard and not a large standing army. He feared a lack of federal support for the National Guard, which his own state had done without until 1927, because he believed that without a National Guard, states would be dependant on the War Department in cases of disorder. He believed that such dependence would lead to the states fading away as political entities.[55] He was the only adjutant general to express this fear so strongly in his reports, and he was an odd spokesman for this fear. Of all states and most territories, "Battle Born" Nevada had the least militia tradition. And despite his fears, or, more likely, because of them, he remained blind to the irony of his position. As with most others who argued for the National Guard from a states' rights position, he never saw any contradiction between his insistence on a strong National Guard to protect the states from federal encroachment and his demand that the federal government assume most of the financial burden for the National Guard, or any other militia the state might create.

While championing of the National Guard in its state role, he never showed much interest in creating a State Guard during the war. After his trip to the combined convention of the Adjutants General Association and the National Guard Association in April, 1943, he noted that "State or Home Guards" were a hot topic among other adjutants general. He realized that his state was one of only four that had not created any such forces but felt that they posed a "large problem" in states that had them, although he never recorded just what "large problem" he felt a State Guard would pose. Despite his misgivings, he advised the governor that "consideration may well be given to the desirability or necessity of authorizing such an organization for the future if any internal protective force is deemed necessary in our own State of Nevada."[56] But despite that advice, he never pushed for its implementation, and so no other force was ever seriously considered.[57]

The remaining state without a State Guard during the war was Montana. Although a large state in area, its population density was among the lowest in the nation. Still, the mining city of Butte had experienced labor violence in the past, and the state had a large number of foreign-born workers. However, the war removed many young men from the state, with an estimated 55,000 men and women serving in federal forces in 1943.[58] Added to this was an exodus of workers to industrial jobs on the west coast, which left the state with little fear of labor unrest within its borders. Under the circumstances, the state worked more on addressing its own labor shortages in a booming economy and placed state military forces on a low priority. The office of the adjutant general remained vacant for much of the war, after two men resigned from the position to enter federal service in 1941 and 1942. Nevertheless, the adjutant general's office continued to oversee what military preparations the state took. Noting that it was one of the four states without a home guard, the adjutant general's office prepared plans for a small home guard should the legislature approve the creation of one. It maintained copies of all training literature and directives on home guards that the National Guard Bureau had sent in case of a change in state policy. The office also kept a file of reports and forms from the home guard of other states for reference and suggestions.[59]

In spite of these preparations, the state legislature opted to depend on a far broader idea of militia as a hedge against disorder or the more unlikely scenario of invasion. Montana had a culture of gun-ownership, and state officials took notice of this when establishing the state's plans for the war. Early in the war, the adjutant general's office instituted a voluntary statewide program of gun registration through the county sheriffs, to give it an idea of "the number and caliber of guns which could be used in case of extreme emergency."[60] The adjutant general's office also kept a card file of members of the Civil Air Patrol and members of rifle clubs affiliated with the National Rifle Association. It reasoned that "knowing the whereabouts and equipment of these men in case of local emergency is deemed very valuable." This program was never intended to become, nor was it ever, more than a list. No records have surfaced of units forming or drilling on their own. With most homes in the state containing a firearm, Montanans believed their state secure during the war. These four states—all in the interior of the nation—along with the District of Columbia and the Canal Zone, never established a militia force to replace their National Guard during World War II.[61]

Despite regional variations, for the majority of states that did create a State Guard, their wartime militia closely resembled their departed National Guard. Although a few states had black National Guard units, most states did not, and black units formed a small minority of the National Guard. States with a history of National Guard units comprised of black troops usually included black units in their State Guard. The "separate" company, battalion, or regiment was the usual means of organizing black units in the National Guard, and this practice carried over into the State Guard. The last of the Ohio National Guard, the Second battalion, 372nd Infantry, entered federal service in March, 1941. In the segregated army, this black battalion was separated from the rest of the Ohio National Guard and regimented with black troops from the National Guards of Massachusetts, New Jersey, Delaware, and the District of Columbia.[62]

The total Ohio State Guard had one brigade comprised of three infantry regiments and two battalions of Ohio State Naval Militia, which replaced the Ohio Naval Militia, after it had been absorbed into the U.S. Navy.[63] However, the State Guard also included the First Separate Company, in Cleveland, which was directly under the adjutant general and not part of any battalion or brigade. The First, and only, Separate Company, was for black State Guardsmen, although this was not made explicit in reports of the adjutant general.[64] The company commander, Captain Charles Gardner, temporarily served as an aide to the president of Liberia from May 30 to June 2, 1943, while the president was the distinguished guest of Ohio.[65] The West Virginia State Guard, in addition to its two regiments, also maintained a separate battalion for black State Guardsmen.[66] The white regiments each had ten companies and a band, while the Separate Battalion had two companies, located in Welch and Charleston, and no battalion headquarters.[67]

But while most states created units that followed the patterns established by the National Guard, some State Guard organizations blazed a few new trails of their own. Foremost among the trailblazers was the State Guard of Puerto Rico. Separated by language and culture, as well as geography, from the rest of the United States, the island territory developed a State Guard in many ways unique in the nation in World War II. The legislature of Puerto Rico authorized a State Guard on May 9, 1941, and appropriated $50,000 for it. However, the act of Congress of October, 1940, which had amended the *NDA* to allow states to create another militia, did not include Puerto Rico. In response to a request from the island's governor, Congress passed another amendment on August 18, 1941, that specifically allowed Puerto Rico to create a home guard to defend and maintain domestic order on the island. The territory also wanted a militia to respond to the hurricanes that periodically swept over the island. Actual organization began in January, 1942, with the territorial legislature appropriating a further $260,000 for the force.[68]

The force it created showed far more creativity and flexibility than those organized in most areas of the country.[69] The Puerto Rico State Guard originally contained five battalions with a total of twenty companies, each with three officers and sixty-two enlisted men. This structure did not satisfy the desire of the Puerto Ricans to organize militia for local protection. In order to take advantage of local enthusiasm in fifty-seven other towns, the governor increased the number of companies in the original State Guard structure to thirty, but added another tier of militia for the remaining towns. The thirty companies, called "full-time," were to receive pay for armory drills. Another fifty-two companies, called "part-time," had members serve for a token fee of a dollar per year.[70] The part-time companies were liable for emergency service only. The island did not have enough arms for all these militia companies, which greatly exceeded what the War Department would support. In order to allow all members to train with weapons, the force reorganized with each battalion having one full-time company, with the others part-time. Each battalion had only enough arms for one company, so arms were shifted between companies for training. By June 30, 1942, Puerto Rico had a total of eighty-two full-time and part-time companies, plus a detachment and band. This gave a total strength of 306 officers and 4,603 enlisted men. The mustering of two companies had been delayed because gasoline rationing had prevented medical officers from examining the men. These companies, which had been training despite the delay in their acceptance into the State Guard, contained another 6 officers and 110 enlisted men.[71]

But aside from the State Guard's large size and anticipation of the "reserve" or "Minute Man" concept, the Puerto Ricans excelled in creating and maintaining ties between State Guard companies and communities. In many respects, Puerto Rico ran ahead of the nation in its creative use of the State Guard to channel enthusiasm and, at the same time, in using the State Guard to maintain that enthusiasm in the communities.[72] As part of its efforts to make each town accept the local State Guard company as its own, every company had a "Godmother," a local woman from the community, and respected men in the community were designated as "honorary officers."[73] As with many states, the Puerto Rico State Guard maintained its own newspaper, the *Centinela Alerta,* which was financed by selling advertising space in the paper. Free copies were sent not only to members but also to civil and military officers of the United States and Latin America. The Puerto Rico State Guard band, considered one of the finest on the island, gave free concerts in public squares at the request of mayors and the governor, and also played on the radio.[74]

As might be expected in a Spanish-speaking part of the United States, acquiring textbooks the State Guardsmen could use remained a problem. Eventually the state staff acquired one thousand copies of the *Manual del Soldado,* which had been published by the Puerto Rico National Guard. Brig. Gen. Luis Raul Esteves, the adjutant general, had written the text, and he quickly approved its release for use by the State Guard. His office also began writing a manual on guerrilla warfare because it could not find any such federal publication, at least none in Spanish. The difference in language forced the Puerto Ricans to create much of their own training material, such as correspondence courses for officers.[75]

By 1943, the basic organization of the Puerto Rico State Guard had been completed, although many problems remained. The lack of federal equipment caused difficulties, but that was true of almost every State Guard in the country. The force did well in federal inspections in spite of a lack of available training camps for the large-scale summer training enjoyed by most states. With the U.S. Army as well as the National Guard away from Puerto Rico, the State Guard had to depend on Great War veterans to assist in training. Federal conscription took its toll on units in smaller villages; two companies disbanded because of a lack of qualified officers after the original officers were inducted into federal service. These companies were in small towns with only part-time companies, and no other potential offices were available.[76]

The immense size of the Puerto Rico State Guard, about five times as large as what the federal government would support, forced it to find sometimes unusual solutions to training difficulties. The entire force of almost five thousand men had only one thousand federal rifles and bayonets. The rifles were redistributed to allow each company to receive enough for one platoon, with the rifles shifted among platoons for maximum training. The shortage of rifles was so severe that, for preliminary instruction of new members, German rifles from the Great War, which had been used as decorations in armories, were taken down and used by recruits for basic weapons drills.[77] Despite problems, improvisation and creativity overcame the geographic, cultural, linguistic, and historical separation of Puerto Rico from the United States, and allowed the island to maintain a large and popular State Guard in a part of the nation outside of the areas assumed to be most supportive of militia.

While Puerto Rico chose to create a far more ambitious State Guard than most states, at least one state chose to look further into the past than the National Guard for its model. In the summer of 1940, Connecticut again called on its First and Second Companies of the Governor's Foot Guard, and its Horse Guard, to provide the cornerstone of a wartime State Guard. These militia units were notified to prepare for field training as State Guard, the first such field training they had undergone since the Great War. The state's plan called for these units to form the Provisional Brigade of Infantry and the Provisional Squadron of Cavalry. The brigade could be expanded into two or more regiments in both of the state's military districts if deemed necessary. The adjutant general, Brig. Gen. R. B. DeLacour, also acted as the brigade commander, with his state military department headquarters doubling as the State Guard headquarters, thus eliminating the need for a regimental headquarters. The army's First Corps Area Commanding General in Boston studied and approved the Connecticut plan. Although the War Department realized that the unique nature of Connecticut's militia system precluded many other states from following its methods of creating a State Guard, it did issue an "official communication . . . offering recommendations for the proper organization of State Guard units throughout the country . . . [that used] for the most part the original Connecticut plan including the elimination of regimental headquarters and the establishment of the Guard in Battalion units."[78]

Organizing the force proved far simpler than arming it. Because Connecticut began the creation of its State Guard in advance of the federal legislation that allowed the War Department to issue arms for state forces, it originally attempted to buy its own rifles. However, the high cost, estimated at $100,000, and the uncertainty of receiving them, led the state to try a different route. At the request of the adjutant general, Representative J. Joseph Smith of Connecticut presented a bill in Congress, which soon became law, that allowed the War Department to loan weapons to states for their State Guards.[79]

Despite some surprise by people whom General DeLacour described as "uninitiated," officers of the Connecticut State Guard were chosen through the old militia tradition of election. The enlisted men elected their own company officers from candidates who had served in the Great War. The adjutant general believed this system to be the most practical because men would elect their natural leaders and said that he had discussed it with the army. He did not, however, record the reaction of the army. After the election, the names of the winners were submitted to the adjutant general, who then recommended to the governor that they be commissioned. Afterwards, promotions followed the normal practice of the National Guard, which was based on recommendations of superiors.[80]

The method of selecting officers was not the only throwback in Connecticut; the State Guard also provided opportunities for horse cavalry, which had been rendered obsolete on modern battlefields. The Governor's Horse Guard, like the Governor's Foot Guard, was a militia organization of long lineage that had remained outside the National Guard. Although horses had no place on modern battlefields besides supply transport, horse cavalry remained useful for patrolling the miles of power lines, pipelines, and the many reservoirs throughout the state. The horse-borne troops proved so efficient at their tasks that the State Guard inducted several more units that had been formed spontaneously by horse owners in the state, units such as the Litchfield Light Horse and Gold's Dragoons of Fairfield. Although the men in these units provided their own horses, the state assumed the burden for feeding the animals when on duty, as well as the liability for damage to the horses.[81]

The governor ordered the State Guard to alert status on the night of December 7, 1941, but in at least two battalions, the men had already assembled on their own initiative and were at their armories awaiting instructions. On that Monday morning, the First Area Commander in Boston requested that each state in his area provide armed sentries at critical points. The Connecticut State Guard accomplished this by noon on Tuesday, December 9. In addition, the Military Police company that had been part of the Headquarters of the State Guard was detached and assigned to the State Police Department. They would spend the next several weeks on constant active duty.[82]

One of the greatest problems in the State Guard was moving men and supplies when called to state active duty, because suitable trucks were unavailable for issue to state forces. Some of the transportation problems of the State Guard were solved by the use of the Connecticut Women's Motor Corps, which provided transportation for officers, especially medical officers. The women in the Motor Corps donated their time and cars, as well as their rations of gas, tires, and oil to the State Guard. During periods when members of the State Guard were on active duty for an emergency, the women of the Motor Corps brought meals to the men. The unit had at least 440 women and their cars.[83]

The adjutant general later reported that he had received no complaints from any State Guardsman who was unhappy with having to perform extended active duty in the days after the attack on Pearl Harbor, but after a week he began to receive requests from employers who wanted him to release individual Guardsmen so that they could return to work. He then ordered the release of men desperately needed at their civilian jobs and for commanders to recruit replacements.[84] The question of the appropriateness of the State Guard for long periods of active service would trouble the state military leaders for the first years of the war. But with war a reality, training took on an added sense of emergency.

After the initial confusion of the months after Pearl Harbor, the State Guard throughout the nation began to grapple with the reality of war. The first major training event for the State Guard came in June, 1942, when the First Services Command, which encompassed the six New England states, conducted a weeklong school on the campus of the Middlesex School, a private high school, in Concord, Massachusetts. Maj. Gen. Sherman Miles, commanding the First Services Command with headquarters in Boston, accompanied by the region's six state adjutants general, went to Concord for the week to observed the training.[85] Some army soldiers from nearby Camp Devens were on hand to assist in training the State Guardsmen. The school was believed to be the first in the nation to teach guerrilla warfare, reflecting the possible employment of the State Guard in the event of enemy raids or invasions.[86] Approximately eight hundred men attended the camp, of whom some seventy-four were officers, five hundred were selected enlisted men, while the remainder came to support the training. The men were formed into four companies for the week of training. Unlike training at the armory, for which State Guardsmen received no pay, enlisted men were paid $1.55 per day, while officers received $2.50, with the money coming from the states. The camp was held on the hottest days of the summer that year, with a particularly grueling five-mile hike. A Catholic and a Protestant chaplain, both of whom had served as army chaplains in the Great War, were on hand to attend to the men's spiritual needs. As with the normal practice for National Guard encampments of the era, the governor of Massachusetts came to review his assembled state troops at Concord that Friday afternoon. But this was no annual summer encampment of the National Guard, like the ones previous Massachusetts governors had visited, nor was it similar to the State Guard camps of 1918 and 1919, which governors had also visited. Instead, this was a camp of instruction for selected State Guardsmen, rather than the whole State Guard, and it included men from all six New England states. The men selected to attend were then expected to take their newly learned skills back to their units and act as a leavening agent to the State Guard as a whole. For the governor, this distinction mattered little, for in the midst of a world war, the chief executive of the commonwealth could review his soldiers, and be photographed doing so, as they trained under the hot sun, all within an hour's drive from Boston.

The State Guardsmen had to learn more basic skills before the more exciting ones. An executive from the Boy Scouts of America came to the camp to instruct the state soldiers on basic camp discipline. The chief of the Worcester police department, Maj. Gen. Thomas F. Foley, who was also the commanding general of the Second Division, Massachusetts State Guard, taught most of the classroom instruction. But classroom instruction provided only part of the lessons taught at camp. The New England State Guardsmen spent much of their week training to fight enemy commandos who came by sea or air into their states. They captured and hauled off for questioning soldiers from Camp Devens pretending to be paratroopers, and ambushed their convoy with practice bombs made out of flour. They practiced using grenades, digging trenches, charging across fields with their shotguns, making incendiary and smoke bombs and Molotov cocktails, and creating roadblocks. Newspaper photos of the event showed men in their late thirties and forties training but apparently enjoying themselves.[87] To add to the realism of the training, a State Guard plane from East Boston flew over the men in Concord and performed a mock strafing while the men on the ground took cover. The men were up and rolling out of their tents at 5:45 A.M., and trained until 9:30 P.M. Part of the evening training included films loaned by the War Department such as "Infantry Hasty Field Fortifications," "German Parachute Troops," and "Use of Cover and Concealment."[88] Standard army chow for the hungry trainees came on trucks from nearby Camp Devens twice a day. In addition, the canteen sold "well iced beer and ale," which proved popular with the men, but no hard liquor. Newspapers covering the event mentioned that General Miles was pleased with the results.[89]

The State Guard school had the complete support and backing of the War Department. Lt. Col. John K. Howard, AUS, commanded the school. He had spent several weeks studying the British Home Guard for the War Department early the previous year, along with H. Wendell Endicott, and they were considered experts on the counter-invasion role of local forces. The school also, however, presented an instructor with a more intimate knowledge of warfare.

A featured guest at the school was Bert Levy, a naturalized American of Canadian birth who had fought in the British Army during the Great War, as a soldier of fortune with the Republicans in the Spanish Civil War, and in several other wars. He had written a book about his experiences, *Guerrilla Fighting,* and was considered an expert on the subject. Before coming to the United States to work with American forces, he had trained British commandos, which led the British to recommend him to the Americans for training Office of Strategic Services (OSS) agents. His employment by the First Services Command in instructing New England State Guardsmen in the finer points of guerrilla warfare reflected General Miles's understanding that the State Guard existed at least in part to oppose enemy invasions or raids on American soil.

The training brought observers from allied nations, who saw firsthand the American version of local defense during the war. Accompanying

General Miles in Concord were officers from China, Britain, and Canada, as well as Col. Natalie Hays of the Massachusetts Women's Defense Corps.[90] A few members of the Women's Defense Corps were involved in the encampment running the commissary and driving trucks in support of the training. Also on hand to observe the training was the military historian Col. R. Ernest Dupuy, then of the War Department Bureau of Public Relations.[91] Colonel Dupuy was on hand no doubt to publicize that Americans throughout the land were trained and ready to defend their communities from enemy attacks.

Despite the activities of many states and territories in the days and weeks after the United States entered the war, not all state adjutants general reacted in a similar manner. Several states had not created any State Guard before the attack on Pearl Harbor, usually fearing that any effort toward creating one would be wasted when the National Guard returned and any State Guard was disbanded. The early days of the war found these states scrambling to organize units. Typical was Louisiana, whose governor did not see any need to create a "Home Guard Organization" before the nation entered the war. He believed that the State Police and the many federal troops in the state would be adequate to deal with any difficulties that local police could not handle. His adjutant general nevertheless had formulated plans for a State Guard should the need arise.[92] After the United States entered the war, the state implemented the plans and maintained a State Guard of "nearly 3,000 members," despite a continuous change in personnel and company commanders.[93]

Although the State Guard of Louisiana resembled the State Guards in most other states, in one aspect it stood alone. The state adjutant general saw little use in employing the State Guard against an invasion force, or in fighting a guerrilla war on occupied U.S. territory. He saw the American State Guard as quite dissimilar to the British Home Guards and abhorred the "over-zealous" reports from Britain and the "extreme nervousness" that prevailed among East Coast and northwestern states. He took a far more conservative view toward the State Guard and, long before most other state military leaders in the nation, foresaw their usefulness only in the normal peacetime missions of the National Guard. In the unlikely event of invasion, the State Guard would be best employed in assisting the evacuation of civilians, and in arresting saboteurs.[94] In this view, he was in the minority, not only among state leaders, but among State Guardsmen as well, who relished the potential for something more romantic than filling sandbags during floods and patrolling streets during riots. As the war progressed, the State Guard throughout the nation slowly became reconciled to the vision of the State Guard held by the adjutant general of Louisiana. It held less of the glamour promised in resisting enemy commandos, but allowed the American State Guard to continue and to thrive after the initial invasion scare faded, and to perform needed functions well into 1947.

The more national scope of the State Guard of World War II made it far more homogenous than the various militia raised for state service during the Great War. Nevertheless, regional variations of population density, apparent danger of invasion, and enthusiasm for militia service created local differences in the State Guard. The Pacific region, fearing that attacks or invasion were inevitable, created large state forces, which often performed long periods of active service. Small, lightly populated states, or states with few or no urban industrial areas usually created much smaller State Guard forces with more limited missions. The Northeast tended to create forces similar to the ones raised in the Great War, while the South sought to create State Guards resembling their peacetime National Guards. Within these broad characterizations, many variations remained.

CHAPTER 8
Assuming a Greater Role

At the end of January, 1943, a woman from the District of Columbia wrote to the War Department trying to clarify something that had been bothering her. She remembered the National Guard entering the army for a year of training, but she did not know what had happened to it when the United States entered the war. She described herself as a "one-woman-rumor-stopping-campaign" but could not answer questions about the identity of the men whom she and people she knew had seen protecting bridges and other important places. She knew enough about the military to notice that, although the uniforms and insignia of these mysterious men were similar to army uniforms, they were nevertheless distinct.[1]

This simple query from a concerned woman highlighted a basic problem of the State Guard—recognition. To the average American, the terms "State Guard" and "National Guard" remained synonymous. What had in reality been the State Guard of World War II became in common memory the National Guard. That the National Guard existed primarily to fight wars has often been misunderstood by the average American, who sees the National Guard more often when it is performing its peacetime state missions of maintaining order, responding to natural disasters, or simply training at the local armory and marching in the Memorial Day parade. With the National Guard in federal service, these local missions of the National Guard were performed by the State Guard, but for most Americans, the differences between the two institutions were minimal.

The State Guard looked so much like the National Guard before it entered federal service that most Americans assumed that they were the same thing. Personal memories of the war years often contain this contradiction. People with family members who had been National Guardsmen before the war knew that their kin entered the army with the National Guard but often, and correctly, saw them as becoming soldiers in the army rather than continuing to be National Guardsmen. Thus when new companies formed and began drilling at the local armory one night a week, most people assumed that this new militia was the National Guard. That it might call itself the State Guard did not make the distinction clear, as that term has often been used informally in referring to the National Guard. This new State Guard had the same armory, same state missions, and often, many of the same members. Local men who had been involved with the National Guard for years—too many years to be eligible for federal service—and who had reputations in the community based on their membership in the National Guard, were again seen marching down Main Street with the local company on Memorial Day. Most Americans would not notice the slight differences in their uniforms, or that their collar brass had "N.C." or "MASS" instead of the "U.S." worn by the National Guard since the Great War.

When World War II ended, and the National Guard reformed, many of the younger members of the State Guard continued to drill at the local armory, only as members of the National Guard instead of the State Guard. Thus many former State Guardsmen themselves have largely forgotten their former status as strictly state soldiers during the war. Instead, individuals who served as State Guardsmen as well as those who did not, erroneously recall incidents from the war years when "National" Guard ambulances brought wounded veterans home, or when "National" Guardsmen patrolled streets after a storm. The institutional memory of the State Guard became grafted to that of the National Guard in the minds of most Americans.[2]

This conflation had been assisted, albeit indirectly, by National Guard Bureau policy during World War II. Maj. Gen. John Williams, the acting chief of the National Guard Bureau, always encouraged National Guardsmen who had been released from active service to join the State Guard. The army had discharged from active service many for a variety of reasons after the initial induction of the National Guard.[3] Throughout the war, more former National Guardsmen would be discharged from the army because of age, physical problems, wounds, and dependency. With the National Guard absorbed into the Army of the United States, a discharged former National Guardsman had no National Guard units under state control to join. However, National Guardsmen released from active service could join their State Guard without jeopardy to their National Guard status. When the National Guard returned to state control, the National Guardsmen who had earlier been released were expected to terminate their membership in the State Guard and rejoin the National Guard if they were eligible.[4] The presence of the recently returned National Guardsmen in State Guard units brought in experienced men and gave these new State Guard units a sense of continuity in their communities. The question remained, however, about the mission of the State Guard during wartime. The question of whether they were simply to perform the normal peacetime missions of the National Guard or to tackle more exciting roles persisted at least into 1944.

On March 20, 1943, the governor and the adjutant general of Vermont traveled to Boston to see Maj. Gen. Sherman Miles of the Army First Services Command. The Vermonters were angry over the lack of federal support for their State Guard. General Miles agreed with his visitors that a gap existed between what the War Department expected the State Guard to do and the equipment that it gave them to do it with. The pair from Vermont saw the primary role of the State Guard as opposing an enemy invasion; the training the army had instituted for state forces, such as the First Services' Military Police school and the State Guard Officers school, reinforced this idea. The governor believed that morale not just in Vermont but the nation would plummet if the role of the State Guard were reduced to one of a second-line state police. General Miles had expected and dreaded the time when a governor would raise this issue because he believed the army was in a false position vis-à-vis the State Guard. It did not equip them for the missions General Miles and others believed it had planned for the State Guard. Troubled by the issues raised by his visitors from Vermont, General Miles wrote to the office of the commanding general of the Army Service Forces that same day. General Miles had enormous sympathy with the State Guards and felt the army should decide what it actually wanted State Guards to do in the event of invasion and properly equip them for the mission.[5]

The reply from the War Department came in less than a week. In it, Maj. Gen. W. D. Styer, chief of staff of Army Service Forces, expressed bewilderment at General Miles's letter. "The War Department never assigned to Service Commanders nor to State Forces the mission of resisting hostile attack." He restated the long-standing War Department policy that responding to external threats was a federal mission. State forces were for internal threats. State forces received some combat training to allow them to respond against commando raids and enemy paratroopers, as well as to cooperate with federal forces in the event of invasion, but this was seen as an extreme situation at most, and not the primary focus or mission for state forces.[6]

The episode underscored a long-standing problem over the mission of home guard forces during wartime. With their nation at war, twentieth-century American men tended to join home guard groups for the same reasons that their ancestors had joined similar wartime militia groups—to defend their homes against invaders. But during the world wars, the War Department saw the role of militia quite differently. It wanted home guards simply to fill the normal, peacetime, state missions of the National Guard until such time as the National Guard would be returned to the states. This role would relieve the War Department of the need to assign federal troops to deal with civil disturbances and natural disasters during wartime. The question over missions began with the creation of the state forces before the induction of the National Guard. The urge to create state forces often preceded any agreement over how those forces would be used. In the months before the United States entered World War II, state leaders often looked back to the spring and early summer of 1917, when the National Guard was on active duty protecting vital points in each state, in framing their intentions for a State Guard in 1940. State leaders also saw the State Guard providing combat assistance to the National Guard or army in the event of invasion, although in a clearly subordinate role, in addition to providing armed patrols for dams, power stations, and critical industries.[7] After other states watched California go broke maintaining what amounted to a standing army after Pearl Harbor, and individual State Guardsmen with families to support realized that state active duty could be a cold and boring ordeal for little money in a booming war economy, states and the War Department agreed that the purpose of militia forces was to respond to acute crises rather than chronic security concerns. After the first months of the war, few on any level wanted the State Guard employed as full-time sentries around the nation, but consensus proved more elusive on the type of acute crises in which the State Guard should be employed.

Most of the State Guardsmen, especially those on the west coast or in reserve or "Minute Men" units, saw their mission as either opposing raids or invasions, with the mundane missions of responding to natural disaster and civil disorder a secondary role. The War Department saw things just the opposite. It placed little or no faith in the utility of state forces in opposing an enemy invasion, and allowed the State Guard to train for such missions more as a morale builder than as a primary mission. The War Department did not want state forces to absorb badly needed equipment for a role to which the War Department believed them ill-fitted, and unlikely to face after the spring of 1942 anyway. The National Guard Bureau explained to the state military leaders that the War Department was not assigning combat missions to State Guards. If service commanders or state officials wanted to give combat missions to state forces, the War Department would not object.[8] However, the War Department reminded states that it would only support the State Guard at a level equal to one-half of the peacetime National Guard. Any forces the states created above that level, be they regular State Guard or "Minute Men," would receive no federal support.

Despite federal and state disagreement over the main role of the State Guard, state military forces performed another valuable function for the War Department. As with its predecessor of the Great War, the State Guard of World War II became an unofficial source for noncommissioned officers for the federal military. With the rapid and massive expansion of the federal military during the war, a man with any prior military experience had a distinct advantage over his fellow inductees. Many former State Guardsmen became NCOs immediately upon reaching federal training camps, or even received appointments to officer candidate schools. In this way the State Guard made a contribution to the federal war effort by providing preinduction training for selectees. Throughout World War II, State Guardsmen were conscripted into the federal military. Most of these were the physically fit, younger men, some of whom entered their state force at sixteen or seventeen years of age, before they became liable for federal conscription.

Halfway through the war, the War Department became curious about the numbers of former State Guardsmen who had entered federal service. In the spring of 1943, it sent a circular letter to all state adjutants general asking how many former members of the State Guard had entered federal service between January 1, 1942, and April 30, 1943; whether they had served in the State Guard as officers or as enlisted men; and how long they had been in the State Guard before entering federal service.[9] By the summer of 1946, the National Guard Bureau counted a loss to conscription of 130,405 State Guard enlisted men and 4,753 officers.[10] Although this hardly represents a sizable portion of the twelve million or so Americans who served in the federal military in World War II, it does represent a significant source of noncommissioned officers for the army, and a great impact on the men themselves. Their experience in the State Guard gave them a head start compared to their peers with no prior military experience. Although the War Department did not count State Guard service for longevity pay, it did recognize the benefits of such training, and most former State Guardsmen quickly rose in rank above inductees with no such experience. The induction of 130,000 or so State Guardsmen brought men who could enter as corporals, sergeants, or officer candidates into the greatly expanded federal forces.

Former state military service of recruits could be desirable to the federal military branches, but the constant hemorrhage of trained State Guardsmen into the federal military placed a burden on state military leaders who needed to maintain a coherent force despite annual turnovers of personnel approaching 100 percent. General Williams emphasized the need for publicity regarding the State Guard in order to make good their massive losses. He estimated that to continue to field a force of 2,000 State Guardsmen, over a three-year period a state would have to recruit 7,163 men because of the large annual turnover. He encouraged the states to publish more hometown news of State Guard activities in their local newspapers.[11] One of the major complaints of the men in the ranks of the State Guard was the lack of recognition for their efforts. To remedy this, the War Department always encouraged state military departments to implement recognition programs. The NGB advised state military departments that they could issue their own ribbons to their State Guardsmen as long as state ribbons did not have designs similar to any federal or foreign medals or ribbons. The NGB suggested that states create ribbons to reward State Guardsmen for perfect attendance at drill for a year or more, small arms proficiency, or a veterans' badge.[12] Although such state awards could not be worn on a federal uniform if a State Guardsman later entered federal service, these tokens offered tangible rewards for service.

Still, the War Department wanted to ensure that the public did not confuse State Guardsmen with members of the federal military forces.

Throughout the war, it continued to be uneasy with State Guardsmen wearing their uniforms in situations not part of their duty. In the summer of 1943, a member of the Maryland State Guard got into trouble for wearing his uniform while he was in Florida. The Marylander was not attending any training or ceremony; he just liked wearing his uniform in public while on a trip. The army did not look kindly on such infractions of civil laws protecting federal uniforms.[13]

If War Department officials felt threatened by State Guardsmen being mistaken for federal soldiers, it also began to see the State Guard as a way to preserve some ancient traditions. Military units have always taken pride in long lineages, and some National Guard units held lineages reaching back to the colonial period. Throughout the war, because of reorganizations, the War Department discharged some former National Guard units with long histories. In practice, this meant that the unit was disbanded and the lineage returned to its home state, while the men in it, many of whom were selectees rather than men who had entered federal service with the unit, went to other units. Some state military departments, unwilling to end long lineages of former militia units, gave those designations to State Guard units to keep the continuity. When the 176th Infantry Regiment, which had formerly been the 1st Infantry, Virginia National Guard, was inactivated on July 10, 1944, companies in the Virginia State Guard got the lineages of the Richmond Light Infantry Blues, Richmond Grays, Petersburg Grays, Huntington Rifles, and Nottoway Grays.[14] Thus the State Guard took on a new, albeit passive, mission: providing continuity in unit lineages.

Acquiring ancient lineages might keep traditions alive, but the War Department had more practical reasons for desiring the states to keep a State Guard. Most of the active duty State Guardsmen around the nation performed was in response to flooding or fires. Occasionally State Guardsmen performed missions more directly related to the war. The Ohio State Guard assisted the State Council of Defense by ensuring that industries complied with blackouts.[15] Three companies of the Maine State Guard used their own vehicles to assist in the search for six German prisoners of war who had escaped while working in the woods.[16] The War Department incorporated the State Guard into its planning for counter fifth column operations, believing that the State Guard could patrol areas not protected by local and state police, and engage enemy airborne troops until the arrival of federal troops.[17]

What the War Department feared most were race riots. Serious questions would soon arise over the ability of the State Guard to maintain domestic order when the vast majority of the federal military left the United States for overseas combat. Some State Guards were able to handle racial violence. Two companies of the Virginia Protective Force spent two days in May, 1943, responding to a call from the mayor of Suffolk over a race riot in the east part of town.[18] The Texas State Guard was able to end violence in Beaumont after white mobs spent the night of June 15, 1943, attacking the homes and businesses of black people. Martial law was declared and the State Guard entered the city the next day, cleared the streets of people, and prevented white mobs from lynching black prisoners in city and county jails. Over twelve hundred State Guardsmen patrolled the city until June 21, when the city reverted to civilian control.[19] Although the Texas State Guard handled the racial strife efficiently, violence in Detroit—heart of the nation's war industries—over racial animosity, housing shortages, and sweltering summer heat, would be more widespread and more worrisome to the nation's military and civil leaders.

Between 1940 and 1942, tension grew between whites and blacks in Detroit, a city where expansion of war industries had created massive housing shortages. In April, 1942, several companies of the Michigan State Troops assisted city police in protecting twenty African-American families that were moving into the federally funded Sojourner Truth Homes.[20] In the next year tensions increased further as the city's population continued to grow, and blacks increasingly chafed at being denied the recreational opportunities afforded to whites. The draft-depleted ranks of the police soon proved too small and too biased to maintain order.

The situation in Detroit exploded in the early hours of June 21, 1943—the day the Texas State Guard turned the city of Beaumont back to civilian control. After an afternoon and evening of minor incidents between blacks and whites in Detroit, rumors spread through the black community of white gangs murdering blacks, and full-scale rioting began. Gov. Harry F. Kelly of Michigan mobilized the State Troops sometime between 10:00 and 11:00 A.M. in response to a request from the mayor of Detroit.[21] At the same time Governor Kelly contacted army authorities in Detroit and Chicago and requested federal troops. However, difficulties over the proper procedure for a governor to request federal troops delayed the arrival of the federal troops until almost midnight. State Troops were no quicker at getting into the city than federal troops. State Guardsmen had to be recalled from civilian occupations, assembled at their armories, and then moved into the city. Mayor Jeffries later testified that when he initially called for the State Troops, only thirty-two men were actually available at the city's Piquette Armory, not the eight hundred he had believed would be, and they had no transportation.[22] Other questions would arise over the legal obstacles in using State Troops to enforce civil law, as well as the proper procedures for requesting federal troops. The most deadly and destructive phase of the riot occurred during the delay in getting soldiers on the streets. Most of the violence and killings ended as the state and federal soldiers took to the streets. For the next several days, six thousand federal troops and more than two thousand State Troops enforced peace. An equal number of State Troops remained on alert in the state's other cities in case the violence spread.[23] Most of the restrictions on the civilian population were lifted after a week, although the sale of bottled liquor remained prohibited. Federal troops remained in the area another week to prevent a recurrence of violence and to train the State Troops.[24] Over four hundred State Troops would remain on active duty in Detroit through the end of the war to protect bridges, tunnels, and, specifically, provide a ready force to civil authorities in case racial violence flared again.[25]

The State Troops augmented the federal soldiers but were unable to restore order on their own. The failure of the Michigan State Troops to prevent violence and quell a riot without federal involvement would make the War Department increasingly uneasy over the reliability of the State Guards. However with manpower and equipment stretched thin in fighting a world war, the federal government would continue to give little material assistance to the State Guards.

By the spring of 1944, the vast citizen army that the U.S. Army had been forming since before Pearl Harbor had almost all been shipped outside of the United States. The bulk had been sent to North Africa, Italy, and Great Britain, where it prepared for the invasion of the continent. Other parts had been sent to the Pacific, where it had begun the long campaign on the road to Japan. As a result, with the continental United States depleted of federal troops, the War Department and state leaders became more concerned with the effectiveness of the State Guards, specifically, their ability to provide internal security. Racial violence in Beaumont, Detroit, Harlem, and Birmingham, Alabama, had shown the potential for unrest in American cities. With the United States stripped of trained federal solders, the War Department dreaded any further internal problems that would be beyond the scope of state governments. It began to query state military departments on the ability of their forces to deal with riots, wanting to know how much training the forces of each state had, and whether each adjutant general believed his force was adequately prepared.[26] By September, 1944, the adjutant general of the state of Washington, in line with most of his peers, reported that all units of the State Guard had received riot control training and he believed that they were fully competent for those missions.[27] The War Department had begun to reissue limited numbers of .30 caliber rifles to the State Guard as early as November, 1943.[28] By that February, the Military Police battalions of the Zone of the Interior, which had been created specifically for crises in the United States, had all deployed overseas. Some officials in the War Department became concerned over the lack of equipment available to the State Guard and urged that some obsolete light armored cars be issued to them, along with communications gear. The War Department empathized with the plight of the State Guardsmen but would not issue any armored cars to state forces because such equipment was not authorized by law. However, the War Department was able to increase its issue of obsolete communications gear for the State Guardsmen's use.[29]

At the 1944 combined meeting of the Adjutants General Association and the National Guard Association, General Williams addressed the assembled state soldiers. He told them that the War Department was becoming more interested in the State Guard as more of the army left training camps in the United States for overseas deployment.[30] With the United States going on the offensive in both Europe and the Pacific, and the training of the mass army reaching a level where more of it could be committed to combat, the nation was being stripped of federal forces. The threat of Japanese invasions or landings by German commandos had become increasingly unlikely, but with so much of federal forces leaving the country, the War Department was becoming concerned over the ability of the states to maintain civil order during crises. The army desperately wanted to avoid bringing soldiers back to the United States to assist states in maintaining order.

General Williams assured the association members that the War Department hoped to provide more support to state forces and saw the greatest need of the state forces as personal equipment, followed by trucks. Williams assured them that the War Department was anxious to assist the states in filling these needs, but it had hit a snag. Although the War

Department wanted the State Guard to receive trucks, the Office of Defense Transportation ruled that federal military vehicles could not be issued to state units. State Guardsmen would have to continue to rely on whatever vehicles they could borrow from their states, or more often, use their private vehicles and continue to use up their gas and tire rations on State Guard activities. On one persistent request from the states pertaining to manpower, the federal government would make no exceptions. Williams explained that although many State Guard officials and state adjutants general had asked him to have State Guardsmen exempt from Selective Service, this was not going to happen.[31] State forces would continue to contain only those not fit for federal service, or those not yet drafted. The enormous turnovers in manpower were a fact of life with which the State Guard would have to deal.

Unlike previous meetings since the start of the war, the conference began to address the postwar role of the National Guard. The world wars had shown the contradiction of the dual nature of the National Guard, which left the states without their National Guard during wartime. The Baltimore meeting would start again the search for a resolution to this dilemma. The National Guard Association recommended a two-tier National Guard in the postwar era. One tier, the National Guard United States (NGUS) would be like the old National Guard. The other tier, the National Guard-State (NG-State) would be for state service only. Both would be funded by the federal government. During peacetime, men could transfer into the state force without a problem. When war came, the NGUS would enter federal service to help fight the nation's war, while the NG-State would remain home in the role filled by the State Guard during World War II.[32] Little came from this proposal, but it did represent the first tentative attempt to create an institutional solution to the contradictions inherent in the state and federal role of the National Guard before the next war began.

Although problems with the State Guard had been discussed at the conferences of the Adjutants General Association and the National Guard Association, the first general conference of state and federal military leaders devoted totally to the State Guard was not held until December 13 and 14, 1944, at the Palmer House in Chicago.[33] Called the State Guard Conference, it was the first meeting of state soldiers during the war that representatives of the Army Service Forces and the National Guard Bureau attended. The conference produced little of substance, but the Army

Service Forces representatives believed it to have been very successful because federal and state military leaders were able to explain their needs and goals and to share ideas. On practical matters, the representatives from the states received detailed instructions on, among other topics, mounting water-cooled machine guns on scout cars, a reflection of the growing concern of using the State Guard in large-scale domestic disturbances.

The states used the opportunity of the conference to request again more material support from the War Department. The main thrust of their position was a desire for more ammunition, instructors, vehicles, and other equipment—all chronic needs of the State Guards. This request the War Department's representatives turned down immediately: no additional support would be forthcoming in the near future.[34] Additional support would eventually come to the State Guard, but the floodgates would not be opened until the end of the war. In the meantime, state military leaders used the resources and creativity of their own states to shape their military forces and prepare their state militaries for the postwar world. In the meantime, the wartime forces they commanded reflected the state and human resources at their disposal, and the threats each state believed it faced.

Although the Alaska Territorial Guard was a vast force, the land was far more vast. Fortunately, the Japanese never landed in the interior of Alaska, only on the distant Aleutian Islands. (With the Utah State Guard, the Alaska Territorial Guard shared the honor of recovering remnants of Japan's balloon attack on the United States late in the war.) The real importance of the Territorial Guard was in creating a sense of shared struggle among the people of a sparsely populated territory. The Territorial Guard would create a new tradition in the Territory that would allow the postwar National Guard to take a more prominent position in Alaska. More than 6,000 men served in the Alaska Territorial Guard during the war in more than 110 units. It stayed in existence for two years after the war and became the basis for the reconstituted Alaska National Guard.[35]

Mississippi provides an example of a more typical existence for the State Guard. Not until 1944 was the state able to supply the State Guard with adequate uniforms, arms, and equipment, but replacement of worn-out equipment remained a problem. By late 1944 and 1945, the federal government had begun supplying the State Guard from stocks rendered obsolete for federal forces.[36] The War Department also supplied the State

Guard with motor vehicles with the provision that the state would provide the maintenance.[37]

Training of the State Guard became more regular as the war years progressed. The officers of the Mississippi State Guard attended a State Guard Officers' School at Fort Benning, Georgia, in May, 1943.[38] Beginning in 1944, the State Guard underwent a week of field training at Camp Shelby, Mississippi, each summer. The state legislature appropriated funds for training and used them to buy equipment; the Mississippians trained without pay. The 1944 encampment, August 6–13, was considered highly beneficial by the army instructors detailed for the training.[39] At the 1945 encampment (August 5–12), the State Guard mustered approximately 2,000 officers and men, with 1,800 present the following year.[40] The last Annual Training was held August 11–18, 1946. In addition to the summer training, the State Guard practiced basic soldier skills, such as first aid and drill, at the company and battalion level the rest of the year, usually in their armories, using training schedules developed with the assistance of the army.[41]

A letter from a company commander, who worked as an attorney in civil life, to his senator, provides insight into the State Guard late in the war. The commander of Headquarters Company, First Regiment, of Cleveland, Mississippi, reported that his company had 128 enlisted men and 5 officers, twice the authorized strength of the average company. His company had the only heavy weapons and communications platoons in the state.[42] He believed that his administration platoon would soon be the only one in the state. He feared that interest in maintaining a State Guard was on the wane but felt the need for state military forces remained high. Most people thought the war would soon be over, and keeping men in the State Guard grew more difficult. But he believed that the European war might not be won in 1945 and that the war with Japan would take another two or three years. He felt that the presence of State Guards helped home front morale because State Guards relieved fighting men, including two of his sons, of worry over possible unrest at home. He said his sons always wrote to him praising the State Guard. Like many people throughout the nation, this company commander wanted President Roosevelt to make some sort of public statement encouraging enlistment in the State Guard. Instead, the War Department encouraged governors to praise their State Guard, further stressing the state nature of the force.[43]

The State Active Duty performed by the force reflected the normal peacetime uses of the Mississippi National Guard. The first state mission came at the end of March, 1944, when the Tombigbee River crested at forty feet above flood stage, killing three people in Aberdeen. The Leaf River near Hattiesburg also spilled over its banks.[44] In response, the governor called out five officers and fifty enlisted men to fill sandbags and aid in rescue operations. The same State Guard unit returned to the Tombigbee area a week later in response to more flooding.[45] The only use of the State Guard in labor troubles occurred at the time the flood waters began to retreat. A dispute at the Ingalls Shipyard in Pascagoula over the use of nonunion workers resulted in the walkout of 750 workers. As a precaution, the governor put the State Guard companies in Laurel, Hattiesburg, and Meridian on State Active Duty, although they remained at their armories. After a day, the situation cooled and the governor released the Guardsmen.[46]

Altogether, the State Guard spent 520 man-days responding to seven separate call-ups for flood emergency duty from 1944 through 1947. Its other principal employment came in response to requests from sheriffs in various counties who needed help to prevent white mobs from lynching black suspects held in custody. The first incident involving the State Guard in civil disturbances came in June, 1944, when a squad of one officer and ten enlisted men was called to the Delta city of Clarksdale to protect a black man on trial for assaulting a white woman. The State Guard ensured that no violence erupted in the tense and racially charged city.[47] State Guards returned to the city to prevent violence in January, 1945, when a black man was charged with killing a white man.[48] A similar case brought out the State Guard on two other occasions, in late 1945 and again in the spring of 1946. In the first week of December, seven officers and sixty-one enlisted men needed machine guns and fixed bayonets to protect the courthouse in Laurel during the trial of a black man accused of raping a white woman. A judicial appeal in the case brought out the State Guard again in October, 1946.[49] The employment of the State Guard in responding to floods and racial tensions mirrored the predominant use of the Mississippi National Guard and demonstrated the dependence many southern states had on their organized militia.

Despite some speculation in the early years of the war that the State Guard might have to act in conjunction with the federal army in repelling an invasion, the actual use of the State Guard proved to be more traditional. As with the Mississippi National Guard before and after World War II, responding to natural disasters and providing assistance to sheriffs in maintaining local order comprised most of the State Active Duty for the State Guard. Although that type of duty was less exciting for the members than fighting enemy invaders, the State Guard nevertheless provided a needed service to the state.

During the peak years of 1944–45, the Mississippi State Guard had an authorized strength of 2,300 men. In practice, the actual muster roles contained between 1,800 and 2,000 men. As the adjutant general stated in his report, the biggest problem with manning the State Guard was that large numbers of men in the State Guard volunteered or were drafted for federal military service.[50] The expansion of war industries and the higher pay and overtime available in manufacturing also induced men to leave the State Guard, which placed a burden on a member's time. To help replace the departed men, the state amended its law pertaining to the State Guard in 1944 to allow boys sixteen years of age to enlist with parental consent. Although many of these younger recruits would later be drafted into federal service, the experience they gained in the State Guard benefited them greatly. Many young men joined the State Guard to gain experience before entering federal service in hopes of entering at a higher rank.[51]

During its existence, the State Guard gave rise to another organization unique to the state, perhaps from its traditional fear of allowing the federal government to have a monopoly of military training. In 1944, Mississippi created a Junior State Guard "with the primary objective of providing military training for the male youths of the state under eighteen years of age."[52] The program acted as a feeder for the State Guard and, later, National Guard, of Mississippi. The adjutant general appointed a board of directors to run the program at the state level. Actual units were formed at high schools, with nine units organized the first year, and fourteen schools participating by the 1946–47 school year. Total enrollment grew from 500 to 869 boys by 1947. The State Military Department made annual tours of inspection of the units.[53]

Rather than a true militia, the Junior State Guard performed a function of the Junior Reserve Officers Training Corp during peacetime. Local units received guidance from an advisory board, with the superintendent of schools acting as head. The State Military Department provided a standardized course of instruction approved by the State Department of Education. Students learned basic military skills, customs of the service, first aid, and marksmanship, and participated in physical training and leadership training. Boys who joined the Junior State Guard received high school credits for their participation.[54]

In his report to the governor just after the war, the new adjutant general, Brig. Gen. William P. Wilson, called the Junior State Guard a "much needed program" and recommended its expansion and continuation "as a source of future National Guard members." He mentioned that should the federal government pass legislation requiring Universal Military Training, then "the State of Mississippi will be far advanced in its pioneering of an established Junior State Guard."[55] At the annual meeting of the Adjutants General Association in 1945, the former adjutant general, General Hays, had presented what became known as the "Mississippi Plan" for universal military training. Early indications were that the federal plan would have eighteen-year-old men on active service for one year. The Mississippi Plan had boys fourteen years and older go through four years of Reserve Army Training Corps, each year of which would count as three months active service—the Junior State Guard program. In the summer of his graduation, a boy would go on active duty for three months instead of the year in the federal plan. General Hays believed his plan would be cheaper and less disruptive, and, perhaps as important, it left the majority of compulsory military training on the state level. General Hays had the backing of educators, unions, and the American Legion. At the time, this was the only state plan for universal military training, which Congress had been debating. The association received the plan favorably.[56] When the federal government opted to forgo universal military training, the Junior State Guard program ended, to be replaced later by the JROTC. Although military programs in high schools could be easily changed, state military forces could not.

In July, 1947, with the war over and the National Guard mustered out of federal service and beginning to reorganize, Congress again changed the National Defense Act to preclude the states from maintaining military forces other than the National Guard in peacetime. The post-World War II dissolution of the State Guard and reorganization of the National Guard proceeded much more smoothly than it had after the Great War, although questions over the role of the postwar National Guard again arose. In late 1947, in its sixth year of existence, the World War II-era Mississippi State Guard reverted to inactive service. They had been retained for two years following the end of the war to give the state an organized militia until Congress settled on a plan for the postwar military and the National Guard was reorganized in 1947. As National Guard units were re-formed, usually in the same locales from where they had entered federal service, the State Guard unit turned over to the National Guard responsibility as the organized militia of the local area and ceased to exist. State Guardsmen who wanted to continue as state soldiers were encouraged to join the reorganized National Guard. For practical purposes the Mississippi State Guard ceased to exist on June 30, 1947.[57] However, few of the State Guard officers met the more stringent physical and educational conditions for commissions in the National Guard. On December 9, 1947, the state of Mississippi promoted most State Guard officers one rank and transferred them to the State Guard Reserve (inactive).[58] The remainder were promoted and transferred on January 5, 1948.[59]

The problem of integrating the State Guard into the postwar National Guard became more of an issue at the 1946 joint conference of the Adjutants General Association and the National Guard Association. Col. R. A. Bailey of Tennessee raised the complaint that federal laws prevented an officer of the State Guard who was otherwise qualified from getting a commission above second lieutenant in the National Guard unless he had served at least six months in the federal forces. State rank would not transfer. A former State Guardsmen who had served for several years and had risen in rank to, for instance, major, would start at the bottom of the officer ranks if accepted as an officer in the National Guard. This provision actually affected few officers, because most men eligible for a commission in the postwar National Guard had been drafted during the war, and most State Guard officers were not young or healthy enough for a commission in the National Guard. Bailey railed against it, drawing much applause from the attendees, who disliked the federal government having much of a role in state forces anyway, be they State Guard or National Guard.[60] Despite their opposition, the adjutants general could do little to change the law because, in effect, the criteria for National Guard officers was grounded in the regulations for the army, and the army was not about to lower the minimum standards for officers. The meeting ended without too much fanfare extended toward the soon-to-be-disbanded State Guard, only passing an "expression of appreciation" to the men who had served.[61]

The end of the war brought various proposals for re-creating the National Guard. As in many states following the Great War, the State Guard of World War II was considered as a base on which to build a new National Guard. In April, 1944, a new proposal for federalizing the State Guard suggested that State Guard units drilling at National Guard armories be given federal recognition as a National Guard company with the word "Provisional" attached. This would give the National Guard Bureau greater control over training and allow for a smoother transition when the regular National Guard unit returned. The idea was that the provisional unit would disband but its members would become part of the regular National Guard unit. The Army Services Forces and the National Guard Bureau declined to support this proposal, citing laws that did not permit such a scheme for the State Guard, and because the War Department, as well as the adjutants general and National Guard Associations, did not want to federalize the State Guard.[62] Many State Guardsmen were ineligible for membership in the National Guard, because many of the enlisted men, and most of the officers, of the State Guard were too old or physically unfit for the National Guard. Such a proposal would bring few new members to a returning National Guard unit.

Despite the ineligibility of most State Guardsmen for membership in the National Guard, many states eyed their State Guard as a base on which to rebuild the postwar National Guard. The most ambitious effort along these lines came from New York, the very heart of the early National Guard movement in the nineteenth century, and the state with traditionally the largest National Guard. The effort was led by Lt. Gen. Hugh A. Drum, who had been commanding general of the Eastern Defense Command and First Army, with headquarters at Governors Island, New York. Originally commissioned in the army at the age of eighteen in 1898, his career straddled the period of reform that brought the army into the modern age. He served in the Philippine Insurrection, Mexican Border, and the Great War, where at the age of thirty-nine he became the chief of staff with the First Army in France. At the time of his retirement from the army on September 30, 1943, he had been its ranking lieutenant general. He claimed that he retired because the centralization of hemispheric defense tasks had left him no role, but in reality he had reached the maximum retirement age of sixty-four. He returned to his home state of New York that October and began his new $10,000 a year, full-time job as commander of the New York Guard.[63] The former New York Guard commander, Maj. Gen. William Ottmann, had resigned the week before to make way for the return of New York's favorite son. Gov. Thomas Dewey made the announcement and Secretary of State Thomas Curran administered the state oath to General Drum at a press conference in Albany. General Drum's mission was to reconstruct the New York Guard to allow it to amalgamate into the National Guard after the war.[64]

General Drum had been a natural choice for the assignment. In addition to his forty-five years of service in the army, he had been the senior instructor of the New York National Guard in 1922 and 1923. A New Yorker himself, his choice was highly popular in the state.[65] Unfortunately for General Drum, having so many old friends among military men of New York soon proved to be a minor burden, as shortly after his appointment had been announced, many people with, in some cases, tenuous links to the general began to ask his help in getting commissions or jobs.[66] In addition to the commander, the New York Guard also maintained full-time positions for two inspector-instructors for each brigade. One was a major, who drew $4,000 per year, the other a sergeant who drew $1,700. Despite this limited, full-time staff, General Drum believed that State Headquarters had too many nonpaid officers.[67]

By January, 1945, General Drum's staff was developing a plan for members of the New York Guard to receive pay from the state for forty drills per year. Like few other states, New York, near the end of the war, was wealthy enough to pay State Guardsmen for training. The same month, the State Guard Association of the State of New York resolved that draft boards should be required to direct all physically fit men deferred from federal conscription to the nearest New York Guard regiment.[68] The state never did draft men into the State Guard, but it could encourage eligible men to join.

In the spring of 1946, Governor Dewey became opposed to the War Department's plan to abolish the State Guard. He believed that constitutionally, the federal government could not put such restrictions on purely state forces. He wanted state forces after the war not to be limited to the National Guard. He recognized that the immense size of New York made it the de facto leader in militia concerns. Like the adjutant general of Mississippi and many other states, he believed that state military forces should be the vehicle for compulsory military service after the war.[69]

When the possibility of universal military training or universal service became increasingly remote after the war, New York began creating a successor for the New York Guard, the State War-Disaster Military Corps. Although this was wholly a state organization, the state wanted it to be federally recognized in a "limited service" capacity and for it to be part of the National Guard system, similar to the "NG-State" proposal of the 1944 Adjutants General Association convention.[70] While many such schemes for a two-tier National Guard system were floated at the end of World War II, New York seems to have been the only state to have actually begun creating such a system. The state would long remain a proponent of the right of states to keep an organized militia in addition to their National Guard.

New York planned to create its postwar National Guard around the organizations and personnel of the State Guard, with the men and officers returning from federal service actively recruited to bring the new National Guard up to strength.[71] The State War-Disaster Military Corps would absorb those members of the New York Guard who were not eligible for service with the National Guard, although General Drum's staff estimated that some 70 percent of State Guardsmen at the end of the war were eligible to join the National Guard. This unusually high percentage reflected Drum's aggressive recruiting of returning veterans into the New York Guard.[72]

Despite the desire of New York and other states to find a role for the State Guard in the postwar military structure, the end of the State Guard was in sight, if not as close as some believed. With the war over, but the National Guard not in existence, many State Guard leaders throughout the nation believed that they would begin receiving vast amounts of surplus equipment from the federal government. In March, 1946, the adjutant general of Indiana tried to get band equipment for his State Guard. He was informed that not only would the War Department not issue band equipment for State Guards but the War Department was not issuing any further equipment to State Guards.[73] Instead, the War Department saw the days of the State Guard as numbered, and awaited the imminent re-creation of the National Guard. Although the army had more war surplus equipment than it needed in 1946, it wanted to issue that equipment directly to the National Guard units then forming rather than issue it to State Guard units, which would then turn it over to the National Guard upon disbandment.

Still, the National Guard would not become an effective force for the states until at least the summer of 1947, and the large number of discharged veterans and the demands of the economic readjustment to peacetime made retaining the State Guard desirable for many states.

Several units of the Iowa State Guard spent the late summer and fall of 1945 restoring and maintaining order after a riot at the State Training School for Boys at Eldora.[74] In the spring of 1946, militia would again prove to be a valuable tool for state government. One of the strongest examples of this came from Virginia in the spring of 1946. By January 1, 1946, the Virginia State Guard retained 309 officers and 3,257 enlisted men organized into fifty-four rifle companies, plus supporting elements.[75] The remaining men were told that they would be discharged on June 30, 1946, but the governor appealed to them to remain until the National Guard again became a viable force. Most of the men stayed, but interest in drilling had clearly waned. However, that spring would see the use of the Virginia State Guard in crushing a strike in the electric power industry, although not in the manner usually associated with militia.

Word that labor and management at the Virginia Electric and Power Company had become "hopelessly in disagreement" reached the governor on March 25, 1946. He promptly put the State Guard on alert. The union threatened to strike on or about March 28. On that date, a conference at the Governor's Mansion outlined the state's plan to deal with this threat to the state's power supply. Present were the adjutant general, assistant adjutant general, the superintendent of the State Police, the commanding general and the executive officer of the State Guard, the commanding officers of seven battalions, two members of the State Corporation Commission, and the attorney general. These sixteen men developed a plan to draft the workers under the state's militia laws and have them assigned to their normal work stations and perform their normal tasks, but as militiamen on active duty.[76] The state justified its need for uninterrupted electricity with the cold weather and with the argument that the strike would affect hospitals and also lead to massive spoilage of frozen foods. To assist the state, Virginia Power and Electric furnished a list of its 2,500 employees. The adjutant general's office then prepared a "Notice of Draft and Order to Report" for each name on the list.[77]

The State Guard entered active status on March 28, and at 9:00 A.M. on March 29, along with the state police, began serving the "Notices" to the employees. Upon being served, each employee became a member of Unorganized Militia on active duty. Each employee was given a copy of the governor's order, with instructions governing subsequent actions. Some employees made "a few derisive remarks," but all complied with the order.[78] The State Guard remained on active duty to ensure that state military law was obeyed by the unwilling militiamen. The plants and facilities were operated through the State Corporation Commission, and the state suffered no interruption of service. The union and Virginia Electric reached an agreement on March 30, and the State Guard stood down. In all, ten battalions participated. On March 30, 1946, the state honorably discharged from active service all members of the Unorganized Militia who had been drafted into state service.[79] This unique show of force by the government of Virginia would be the last major employment of the Virginia State Guard. As the National Guard began re-forming, State Guard companies mustered out. The ancient lineages carried by the State Guard during the war were transferred back to the National Guard. The last of Virginia's wartime State Guard units disbanded June 30, 1947.[80]

In the first half of 1947, most other states also disbanded their State Guard. Typical was the experience of Delaware, which disbanded its State Guard in January, 1947. As was the practice in most states, mustering out came with a public ceremony in the state capital, which included a parade, the awarding of decorations, and a retirement of the colors.[81] During its existence, the Delaware State Guard trained in military skills, marched in a few parades, but unlike the State Guard in almost all states, performed no state emergency services.[82] Through its presence, it gave many Delawareans an opportunity to enjoy a taste of military life without leaving their state. It allowed the traditions of the National Guard to continue while the National Guard was in federal service, and perhaps its existence kept potential trouble from becoming real trouble. But Delaware, a small state with a mostly rural area and one major city, passed the war years without a serious breakdown in order.

As the National Guard began reforming, many former State Guardsmen left their states' military service for good. The higher standards for membership in the National Guard, and the passing of the wartime crisis, left many former State Guardsmen without a role to play in their state militia. At the same time, their former state service had already begun to slip from public memory. Some former State Guardsmen were angry that while part-time civilian components of the federal military, such as the Coast Guard Temporary Reserve, the Woman's Auxiliary Army Corps, as well as draft board officials, got the American Theater or Victory Ribbon, State Guardsmen got neither those nor the American Defense Medal and the World War II Victory Medal. The former State Guardsmen argued that drafted federal soldiers who worked in offices and never left the United

States got these medals, so State Guardsmen should also. The War Department would never budge on this point. Throughout the existence of the State Guard, the War Department consistently stressed their state nature, yet many State Guardsmen strongly desired federal recognition for their service. They had served their states and communities, and indirectly the federal government, for so long and believed that they had been denied any expressions of gratitude or even recognition. Throughout the periods of rationing, State Guardsmen used their personal allotments for gasoline and tires to attend weekly drills and other State Guard functions, and the federal government seemed to ignore their service and sacrifice. State adjutants general even threw their weight behind the request for a federal token of recognition for service in the State Guard. Most of them saw some type of federal ribbon for service in State Guard as appropriate, a ribbon that State Guardsmen of every state could wear, and that former State Guardsmen who later entered federal service could also wear on their federal military uniform to show that they had once served in a State Guard. But the War Department believed that this would set a bad precedent owing to the state nature of the State Guards, and that the navy might not appreciate the army authorizing federal ribbons for state service.[83] No State Guard ribbon was ever issued by the federal government.

Former State Guardsmen of World War II often felt cheated by their status in the postwar era, but they were not the first state soldiers to realize that state service simply was not recognized by the federal government. Capt. George A. Craig had served in the Massachusetts State Guard during the Great War, where he took part in relief efforts during the influenza epidemic and the Boston Police Strike of 1919. Thereafter, he had lobbied long, hard, and unsuccessfully in his quest for a tangible reward for such service. In 1931 he published a history of the Massachusetts State Guard and its Veterans' Association in the hope that men such as he would gain national recognition and tangible rewards for their service.[84] As late as 1947, he was still writing to his senator inquiring if any legislation had been passed giving former State Guardsmen preference in obtaining jobs through the Civil Service, or whether they would be permitted to use Veterans Administration hospitals. He hoped that the enlarged scope of the State Guards of World War II would bring recognition for state service he believed their due. The answer he received was the one he had been receiving since the State Guard of the Great War had been disbanded—No.[85] The only compensation for State Guardsmen would have to come from the states, and in the excitement of returning soldiers from the war, the service of the State Guardsmen in either of the world wars often looked pale.

The State Guard had served their states and communities for as long as seven years and often without pay. Their future depended on the question of the role to be played by the National Guard in the next war. As the National Guard Association began to rebuild strength, it found the dual federal and state nature of the National Guard increasingly called into question.

CHAPTER 9

The State Guard in the Cold War

Throughout the latter half of 1946 and into 1947, the vast bulk of the State Guard disbanded. In thousands of towns and cities across the nation, ceremonies small and large announced the end of units. They had survived as long as seven years, battling floods, fires, and riots. They had searched for fugitives, enemy prisoners, and lost children. Units endured despite annual losses in men approaching 100 percent. They had preserved for the National Guard the traditional roles of militia in states and communities. But the State Guard could not survive the return of peace and the National Guard. By the end of 1947, very little of the State Guard existed, although the future of the National Guard remained uncertain. As in the years following the Great War, after World War II the National Guard had to struggle for reorganization and survival. Many Americans, both in and out of government, began to question the role of a state-based military force in the nuclear age. The National Guard was attacked by military planners who questioned the need for large ground forces in the atomic age and by others who questioned the utility of a military force that would require additional training after mobilization. Within the new Department of the Army was a strong desire to downgrade the National Guard to a strictly state mission, while others, such as the Department of Defense Committee on Civilian Components, commonly known as the Gray Board, wanted to place the National Guard completely under federal control by merging it into the Organized Reserves.[1] The newly created Department of the Air Force planned to sever all connections between the Air National Guard and the states.

The National Guard Association, the main lobby for the National Guard, strongly opposed any such changes. Since the late nineteenth century, the National Guard had sought to be a state-based military force with a federal role as a reserve of the United States Army. State governments tended to support the National Guard's position, both because of their need for a state military force and from a lingering belief in the National Guard as a bulwark of states' rights. When the governor of Mississippi, in an article in the *Army-Navy Journal,* strongly stated his opposition to the Gray Board's proposal, he brought up the usual arguments of states' rights and federal tyranny.[2] Perhaps more important, the integration of the National Guard into the Army Reserve and Air Force Reserve would shift the entire financial burden for maintaining an organized militia to the states. The loss of peacetime state missions for the National Guard would mean that the State Guard, or some other state defense force, would have to be created and maintained wholly with state funds if a state were to have an organized militia. Under the status quo, states had the privilege of maintaining an organized militia in the form of the National Guard largely trained and equipped by the federal government, which the states paid only when on state active duty. With the complete hostility of the National Guard Association and state governments to any of the proposed changes, the dual role of the National Guard, with state missions in peacetime and federal missions in wartime, remained intact.

The world wars had made clear the inherent contradictions in the National Guard's dual role. A parallel debate centered on the future development of State Guard forces during periods when the National Guard could not fulfill its state role because of federal service. The National Guard Association wanted to create a force to fulfill state missions during wartime in order to preserve the National Guard's combat role; otherwise states might clamor to retain control of the National Guard in the next war. The association sought, however, to prevent any such force from taking the National Guard's peacetime role. The National Guard Bureau suggested the creation of a cadre force called the "National Guard Internal Security Force," which would be separate from the regular National Guard of the state but would remain under the control of each state adjutant general, and therefore under the National Guard.

The Operations and Training Division of the Army General Staff disagreed with this plan. Instead, it defined the state use of National Guard troops as more of a police than a military function, and proposed that in the event of mobilization of a state's National Guard, the state increase its state police forces. Any proposal that was based on an expansion of police forces missed the whole point: many members of communities wanted a militia role during wartime. The State Guard movement, as well as the National Guard and the Volunteers before them, existed because of this locally based desire to band together and serve community or nation in times of war. Also, police forces needed to be paid full-time, not only during crises. As a compromise, the National Guard Bureau proposed a plan for wartime that included an expanded state police coupled with a State Guard force with a military police Table of Organization and Equipment.[3] The outbreak of war in Korea caused the debate to be shelved. Instead, the Department of the Army updated its regulations for State Guards from World War II to emphasize company and battalion level forces for internal security, although equipment for such units was not specified.[4]

At the start of the war in Korea, the federal government inducted 8 of the 27 National Guard divisions, 3 National Guard regimental combat teams, and another 714 company-sized National Guard units.[5] Eventually, one-third of the Army National Guard and three-quarters of the Air National Guard would be brought into federal service during the war, but in the late summer and early fall of 1950, the final extent of the participation of the National Guard remained unknowable.[6] As a response to the uncertainty, Congress again changed the law to grant states the right to form a State Guard.[7] Unlike previous amendments, this one included a provision that the authority to maintain these forces would automatically expire in two years if Congress did not vote to extend the law.

The amendment passed only over the objections of the Department of the Army, which still hoped to convert the National Guard into a purely state force and did not want the State Guard to be a standard wartime institution. The strong supporters of the National Guard on the Senate Armed Services Committee overruled the objections of the Army and reported favorably on the bill to the full Senate, which greatly increased its chances of passing.[8] Still, that the amendment passed at all showed that at the start of the Korean War, the State Guard had become the standard response to the loss of the National Guard during wartime, at least for the National Guard, the state governments, and the Congress. This response was not surprising, for the State Guard of World War II had existed for as long as seven years in some states and had only been disbanded for about three years when the Korean War began. State military staffs and many members of Congress simply assumed that the call-up of the

National Guard meant the re-creation of a State Guard. In all, at least half of all states would either begin planning for a State Guard or create at least a cadre of a State Guard.

Unlike with the start of the world wars, the Korean War saw no spontaneous creation of militia forces. The remote war did not inspire the same grassroots need to organize that normally swept the nation at the start of a war. Instead, the impetus for a State Guard during the Korean War came from state military departments. For a few states, notably Texas, California, and New York, the framework for a State Guard had been laid in the years between the end of World War II and the start of the Korean War. California had completely disbanded its State Guard from World War II by August 1, 1946, after the state repealed the *California State Guard Act of 1943*. Former members who applied were transferred to a reserve list.[9] Although this reserve existed as an inactive component of the state's military structure—more as a list of names than a functioning militia—at least some members must have taken their membership a bit more seriously than the state wanted. Less than a year after the disbandment of the State Guard and the creation of the reserve, the state military department needed to publish regulations regarding the wearing of the State Guard uniform by the state reservists, indicating that some members were still wearing their State Guard uniforms.[10] This was the first postwar directive concerning state military reserve forces in the state.[11] Significantly, it referred to the state reserve by a new name, the California National Guard Reserve, which would later be used by at least one other state—Oregon. The new name stressed the subordinate position of the force vis-à-vis the National Guard: the force was to be the state's reserve force in case it lost the National Guard. However, this name was not without its ambiguities. The term had occasionally been used to refer to the Inactive National Guard, which was made up of members of the National Guard who no longer attended drills or annual training but were still liable for active service in an emergency. In addition, the rise of the fully manned units in the Army Reserve and Air Force Reserve after World War II led to the use of the phrase "National Guard and Reserve" when talking about both reserve components. Perhaps because of the ambiguity, the name of California's State Guard of the Korean War era would undergo more changes over the next few years.

The frequent name changes underscored another, more fundamental problem with State Guards after World War II: an uncertainty over their potential missions. Federal, state, and local military leaders had reached the consensus that the State Guard would perform the normal peacetime missions of the National Guard only late in World War II, after any threat of invasion had passed. The war-weariness of the American public, coupled with the dawning awareness of the Soviet Union as an antagonist rather than ally, ended the consensus on the role of state forces in war. In the Cold War, state military leaders would constantly change their conception of their wartime needs, from responding to floods to suppressing Communist sympathizers to evacuating cities before nuclear attacks. No national consensus was found, and the constant name changes reflected the uncertainty of state military leaders.

In December, 1949, the California adjutant general issued another circular concerning the retired and reserve lists.[12] It provided for the creation of reserve areas, reserve units, and attachment of the personnel on the reserve lists to units, as well as more directives on wearing the uniform. The National Guard Reserve was still a paper force, but at least the paper had been organized into specific units and areas of responsibilities for the names on the lists.[13] During the next year, state military leaders again changed the name of the force to the "California Defense and Security Corps."[14] Still, despite the interest of the state in creating a backup force for its National Guard, most of the structure was no more than a paper force, and even as a paper force it was mostly theoretical—no definite plans for actually creating units of real flesh-and-blood men had yet been formulated. That situation changed in October, 1949, when a new state law called for the State Guard to be created when any part of the California National Guard was in federal service, or when State Guard-type forces were authorized by Congress. Earlier state laws only allowed a State Guard to be created when the entire National Guard entered federal service.[15] California's advance planning to create the framework of a State Guard would pay dividends within a year, although nothing indicated that the state saw any particular reason to expect the loss of any of their National Guard in the immediate future.[16]

With the outbreak of war in Korea, and the induction of the National Guard a definite possibility, Governor Earl Warren on July 20, 1950, directed his adjutant general, Maj. Gen. C. D. O'Sullivan, to create the State Guard. As one of the first actions to make it a reality, the adjutant general appointed a former adjutant general of the state, Maj. Gen. R. E. Mittelstaedt, as commanding general, Second Division. To command the

First Division, he appointed Brig. Gen. Ivan Foster, retired, a former regular officer who had been an instructor with the Fortieth Division, one of the state's two National Guard divisions.[17] The planning begun in the late 1940s began to pay dividends in the ability of the state to create a new organized militia quickly after the start of the new war. On August 16, less than a month after the governor directed the creation of the State Guard, the first units of the California National Guard began to enter federal service.[18]

Despite the departure of some of the state's National Guard for federal service, the next adjutant general, Maj. Gen. Earle M. Jones, warned his new state forces not to become too eager to assume a martial presence in the state. Members were not to perform active military duty unless expressly directed to do so by him. The TOE of the force provided for two division headquarters, eight group headquarters, twenty-four battalion headquarters, with each battalion containing three or more companies.[19] The recruiting of the force began on August 15, 1950, although Congress did not amend the *National Defense Act* until September 27, 1950.[20] Each of the two divisions was authorized five officers and eighteen other personnel full-time, in addition to a training officer from headquarters. Clearly the state military department took seriously the possibility of the entire National Guard leaving the state.

On May 10, 1951, the force reverted to the name "California National Guard Reserve" (CNGR) and a week later reorganized along the lines of the Department of the Army's TOE number 19-56 (State), which was the army's plans for an Internal Security Battalion and an Internal Security Company. Under these new tables, each division was to contain sixty-five units, organized into four groups, twelve battalions, and forty-eight companies. TOE 19-56 was the Army's latest attempt to provide a standard plan for State Guards. The term "Internal Security" reflected the Department of the Army's desire to emphasize the state police nature of state military forces. Adoption of the tables by the states was purely voluntary, but lacking any other central plan, most states adopted at least the name for their State Guard. The TOEs provided for full as well as cadre manning. A full division was to have 6,794 personnel, while a cadre division had only 1,815 personnel.[21] The cadre organizations brought the expandable army concept of Maj. Gen. Emory Upton, the late-nineteenth-century American military theorist, to its most drastic application. The U.S. Army never fully adopted Upton's plan, but it proved more practical for State Guards in peacetime. At cadre strength, the division was practically useless as an instrument of the state, yet it created and preserved the framework of a military force, needing volunteers to fill its ranks during an emergency.

During the Korean War, the CNGR continued to expand. The induction of the Fortieth Infantry Division, the National Guard division from southern California, left many areas in the state with no National Guard. To ensure that all areas of the state would have the protection of state military forces in the event of natural disaster or a widened war, the governor authorized the First Division of the California National Guard Reserve, with its divisional headquarters in Los Angeles, to expand to its full TOE strength. The Second Division, based in San Francisco, remained at cadre strength.[22] By September, 1951, the total strength of the men actually drilling as part of the CNGR climbed to 2,799, with 1,908 of those men in the First Division, 866 in the Second, and the remaining 25 at the state headquarters in Sacramento.

Like other states that created a State Guard during the Korean War, California constantly expected federal support to assist the state in arming and equipping its forces, but that assistance never arrived. The state ended up paying over $1,500,000 between 1950 and 1952 for uniforms, textbooks, manuals, and administrative supplies. In spite of this outlay, many of the force's members found that they had to provide their own equipment if they were to have any available for training. The training schedule resembled that of the National Guard for the period. Units were expected to hold forty-eight drills lasting two hours each per year. Units also participated in "week-end active duty schools" that allowed the command to train for a few days under field conditions. In addition, leaders were expected to conduct Command Post Exercises (CPXs) to train headquarters units in handling large problems. Certain personnel of the National Guard Reserve were allowed to attend Annual Training with National Guard units that had remained in state service, and attend army courses at Fort MacArthur.[23] Although the situation did not appear as threatening to California as the first days of World War II, the state nevertheless had earnestly responded with a partial re-creation of its State Guard, and prepared to increase its readiness in case of a long absence of the Fortieth National Guard Infantry Division.

California's neighbor to the north took a more tentative approach to re-creating a State Guard, which was similar to at least six other states that also created no more than a cadre force. In view of the possible induction of his entire National Guard, the adjutant general of Oregon took steps to organize a State Guard, which he christened, most likely influenced by California, the Oregon National Guard Reserve. Unlike California, Oregon stuck with the name "National Guard Reserve" for its replacement militia. Using mostly members who had served in the World War II State Guard, the adjutant general was soon able to complete the cadre staffing of three regiments under the command of a brigadier general. The force included an air wing, also manned at cadre strength, to work as part of the state's air search and rescue program.

The fledgling National Guard Reserve found supplies and home stations difficult to obtain. Bearing in mind that the Korean War might not draw into service the entire Oregon National Guard, the adjutant general kept the force at cadre strength and decided not to intensify recruitment until an actual need existed. Still, he desired federal assistance for the force.[24] He believed that the skeleton units, with members drilling without pay, had prepared themselves well for assuming the National Guard's state mission, but the force was never used. The Oregon National Guard Reserve, like most State Guard forces of the Korean War, never developed beyond the cadre stage.[25] Because of the withdrawal of federal authority for such forces, the adjutant general deactivated all of the National Guard Reserve, except for the headquarters detachment, on September 27, 1952.[26] This remaining detachment consisted of a small staff that was retained for records processing.[27]

Unlike California, Oregon and most other states took few concrete steps toward creating a State Guard before the Korean War. But California was not alone in its attempts to ensure it always had an organized militia at its disposal. After the end of federal authority for the State Guard of World War II, Texas created a State Guard Reserve Corps. Like the State Guard in California between 1947 and Korea, the new Texas force existed only on paper as a hedge against any future inductions of the Texas National Guard. The Reserve Corps contained thirty-six Internal Security Battalions, each with four companies. The battalions were organized into three brigades, and the whole force constituted a single division. In spite of this impressive-sounding structure, like the California National Guard Reserve, the Texas State Guard Reserve Corps consisted mostly of lists of names and organizational charts. A few enlisted men had specific units to which they were assigned in the event of mobilization, while the bulk formed the Enlisted Reserve Pool, to be assigned to units when the need arose.[28]

Throughout the period between the disbandment of the State Guard of World War II and the outbreak of the Korean War, the Texas State Guard existed more as a plan than as a reality. Still, the planning was taken seriously by state military authorities. They developed a training plan to be used in the event of a mobilization of the force. It included such basic military subjects as first aid, unit security, map reading, and drill. However, perhaps in anticipation of its real world role, tactics and techniques for the suppression of domestic disturbances formed a large part of the training plan. State military leaders had learned from World War II that state military forces were unlikely to fight against enemy amphibious or airborne attacks, and instead prepared to respond again to race riots and labor strife.

After the outbreak of war, the Reserve Corps intensified its planning and began activating units. By August 31, 1950, the Reserve Corps contained 1,293 officers and 1,059 enlisted men in its units.[29] The numbers make clear that, although the Reserve Corps was no longer a paper force, it was still a cadre force, perhaps because Congress had not yet authorized the State Guard in federal law. When Congress did change the law to allow new organized militia, Texas, like other states that had begun recreating a State Guard, expected a flood of federal support for these forces. That flood never came, but Texas maintained its Reserve Corps throughout the war and beyond. Although most of the Texas National Guard remained under state control throughout the war, the commanding general of the State Guard authorized Reserve Corps units to recruit to strength in cities that had no National Guard.

The limited nature of the war in Korea had left states uncertain of their response. World War II had seemingly begun a new paradigm: the National Guard entered federal service as a whole; and a State Guard took its place as organized militia. Federal law had guaranteed the National Guard in its entirety would be taken into federal service during war, in advance of reserves or draftees, but Congress suspended this right of the National Guard at the start of the Korean War. The army desired the suspension to allow it to use individual draftees to fill slots in understrength active army units, and to prevent Korea from absorbing all of America's reserve forces, fearing the war as a feint before a main Communist thrust into Western Europe. The Department of the Army and the Department of the Air Force absorbed only part of the National Guard and within a year began returning men and units to state control as drafted soldiers began to fill out the army. As a result, states that in the late 1940s had laid plans for the next war found themselves with plans ill-suited to reality. Perhaps the best example of a state trying to prepare for the future using the plans of the past and the uncertainty of the present was New York.

New York also had more ambitious plans than most states for its New York Guard during the Korean War. In July, 1950, when the first units of National Guard were inducted, plans were begun to create a cadre of the New York Guard. The state's constitution required that the state keep an organized militia, including National Guard, naval militia, or State Guard, of not less than 10,000 enlisted men available to the governor.[30] The New York plan called for an Internal Security Company of the New York Guard to be formed at full strength in each of the vacated armories as each National Guard company left for federal service. A cadre company of the New York Guard would also be formed at all other armories in the state as a skeleton force for creating a full State Guard company if the local National Guard company were mobilized. The plan called for the force to be manned by volunteers between 17 and 64 years of age. The complete TOE of the New York Guard called for the creation of 41 Internal Security Battalions, with a full compliment of 136 companies. Of the 41 battalions, 24 were to be created in the southern area, nearer to New York City, and the remaining 17 battalions in the northern part of the state.[31]

At full strength, the New York Guard would have had 23,806 members. Unfortunately, the state only had enough equipment and uniforms for about 5,000 Guardsmen. In addition, the state had no weapons of its own for the force and worked diligently to borrow some from the federal government. The federal government, for its part, had promised in September, 1950, to provide equipment for purely state forces.[32] In reality, the Department of Defense was hard-pressed to provide properly for the forces it was sending to Korea, and providing for state forces again fell to a very low priority. The shortage of equipment in the army at the start of the Korean War reached such levels that the Department of the Army was forced to recall hundreds of tanks, thousands of trucks, and an enormous amount of other equipment, from National Guard units that had not entered federal service.[33] This acute shortage at the federal level meant that assistance to state forces was many months, or even years, away. Lack of federal material assistance led state military authorities to slow the pace of activating the New York Guard, but companies were formed, and the state bought gear of its own.

By the end of 1951, by which time the federal government had greatly reduced the pace of inducting National Guard units, the state still had some 18,000 National Guardsmen under state control. Despite the ambitious plan for the re-creation of the New York Guard, the state's constitutional imperative that it maintain a force of at least 10,000 men did not yet mandate the creation of the force. The formation of the New York Guard continued, albeit on a smaller scale for the foreseeable future. From the original plan, thirty-four cadre battalions, with a total of 116 companies, had actually been formed, a total of about 700 officers and another 600 key enlisted men.[34] Most of these men were veterans of World War II. Despite the evidence that the Korean War was not to be a general war and that the state would retain the bulk of its National Guard, the adjutant general still hoped to expand the New York Guard to 10,000 men immediately. Both the New York Guard and the New York National Guard that remained in state service were slated to help Civil Defense authorities in the event of an attack.[35] Obviously the adjutant general feared that the Korean War was possibly part of a much larger plan of aggression against the United States.

By the fall of 1951, the general consensus among state military authorities in New York and much of the rest of the nation was that federal law was going to be permanently revised to allow for a cadre of the State Guard at all times, whether or not the National Guard was in federal service. Based on this belief, the states of New York and New Jersey entered into an interstate compact for military aid in time of emergency; Pennsylvania also sought to join the compact. In addition, the governors of Connecticut, Massachusetts, and Vermont had expressed interest in joining the compact.[36] However, interstate compacts required the consent of Congress.[37] Therefore, at the next convention of the National Guard Association, held in October, 1951, in Washington, D.C., the New York delegation pushed through a resolution that placed the association behind a bill in each house that would grant the needed consent.[38] The bills passed both houses of Congress, and became law on July 1, 1952.[39]

New York's victory was short-lived, as federal authority for a State Guard had only about two months left before it expired, and Congress had passed no legislation to allow a permanent cadre of the State Guard. The New York delegation to the National Guard Association tried at the next convention, held in October, 1952, to put the lobbying power of the association behind congressional bills that would extend the authority of the states to maintain a State Guard until the end of 1954. These bills had been introduced by members of Congress from New York at the behest of the New York National Guard.[40] However, as the army had already announced that it would be immediately releasing all antiaircraft artillery of the National Guard back to their states, and that the eight National Guard divisions then on active duty would be released far in advance of the two years for which they had originally been inducted, and that the army would not be mobilizing any further National Guard units in response to the war in Korea, the bills were shelved. Congress allowed the original two-year authorization for the State Guard to expire instead of extending its authority, which would have allowed states that so chose to keep a cadre force. Congress instead indicated that in the event of another mobilization of the National Guard, it would again amend the law to allow states to organize a State Guard. The National Guard Association allowed the matter to drop for the time being.[41] By 1954, in light of the fact that "federal law precludes the organization of State Guards while the National Guard is still in State Service," the bulk of the New York Guard of the Korean War was disbanded. However, in spite of federal law, some of the officers from the State Reserve List remained as a cadre staff to prepare plans in case of future activations.[42]

Only a small part of the New York National Guard had been inducted for federal service during the Korean War, and so the state remained in control of the bulk of its National Guard. However, the Department of Defense did not mobilize units evenly across the nation, and some states lost a far greater proportion of their National Guard. In Mississippi, the federal government inducted 95 percent of the Army National Guard.[43] However, some National Guard units remained under state control and so even with its proportionately greater loss, the state still had no pressing need to re-create an active State Guard.[44] In response to questions of local security and the army's desire for State Guards to remain in either a planning or a cadre stage, the Department of Defense allowed any state that had lost at least 40 percent of its National Guard to organize temporary National Guard units, which could perform state missions and absorb returning National Guardsmen.[45] These new units, designated as "carrier battalions," were to be abolished as National Guard units returned to state control. Mississippi received permission to organize twelve of these type of National Guard units, which were trained and equipped by the federal government, in 1952.[46]

State leaders in Mississippi were not completely mollified by the new National Guard units. Their State Guard of World War II had not been ready to perform state service until 1944, and they did not want to be caught short again. In the event that the war in Korea widened, necessitating the federalization of the remainder of the National Guard, the office of the adjutant general of Mississippi created a mostly paper force called the State Guard Reserve, with one regiment in each federal court district and having a total paper strength of 1,500 officers and men.[47] At least one company, in Jackson, had at least partial manning. The majority of the officers of the State Guard Reserve had been in the State Guard during World War II.[48]

The Mississippi State Guard Reserve of the Korean War was meant to be used in a contingency that never occurred. However, it provided some of the few recorded examples of State Guard troops on state active duty in the United States during the Korean War. The first state active duty came on the night of March 31 to April 1, 1951, when four officers and twenty enlisted men of the Internal Security Company moved north fifty miles from their home station in Jackson to the Durant and Lexington area to "assist civil authorities of Holmes County in enforcing the laws of the state."[49] The enforcement consisted of raiding twenty locations conducting illegal gambling.[50] A similar event occurred on the afternoon of May 19, 1951, when twelve State Guardsmen and thirty-three National Guardsmen raided liquor establishments in Simpson County and seized beer, wine, and whiskey, as well as two slot machines.[51] This use of state military forces for enforcement of liquor laws was not unique to the Korean War era, as the state periodically had used the National Guard for similar purposes in the past.[52]

Not counting state active duty for court appearances stemming from the raids, members of the Internal Security Company of the Mississippi State Guard were on state duty on five more occasions, with the last in September, 1952. All were for enforcement of liquor laws except the call-up of April 30, 1952, when State and National Guardsmen were sent to the BVD clothing factory in the Pascagoula-Moss Point area on the coast. The occasion was a labor dispute and the local sheriff feared violence.[53]

After that, only National Guardsmen appear in the records of the antiliquor raids, which continued throughout the remainder of the year.[54]

Even on the raids that included State Guardsmen, National Guardsmen usually provided the bulk of the manpower. The first raid with the State Guard included the greatest number of State Guardsmen. After that, one raid employed twelve, and the rest usually four. This begs the question of why State Guardsmen were used at all. Perhaps the answer can be found in the background of the officers of the State Guard called to state active duty. One captain in the State Guard was related to the National Guard officer who led most of the raids.[55] The background of a first lieutenant in the State Guard who went on every state active duty the State Guard performed in Mississippi suggests a better explanation. He had been discharged from the army in the spring of 1944 following a motorcycle accident that had left him with a stiff leg. As such, he was ineligible for service in the National Guard, but he could be commissioned in the State Guard. As his whole career had been in law enforcement, and he had attended training at the FBI National Academy and at the Mississippi Patrol School, he seemed a natural choice to participate in the raids. His status as a State Guardsman allowed him to do so.[56] The National Guard officer who led the raids did so in his civilian capacity as head of the Mississippi Highway Patrol. By employing the state soldiers, he was able to beef up his raiding parties and hit several establishments at once. The existence of the State Guard company in Jackson gave him a greater pool to draw from and enabled him to employ men who otherwise would have been ineligible for the raids.

Despite Mississippi's creative use of their State Guard, most states adopted a wait-and-see attitude toward the war in Korea. The threat of an expanded war in which the Soviet Union would play a direct role remained a possibility, but most Americans did not believe that the North Koreans posed a direct threat to their communities. For the dozen states that did begin to create at least a cadre State Guard, the process had a common pattern: the State Guard began as an echo of the World War II model, often with many of the same personnel, but quickly faded away as federal authorization and local concern ended. More common was the reaction of at least twelve other states, like Florida, which laid the foundation for a State Guard but never actually created one.[57] Florida state law authorized the Adjutant General's Office to keep plans current for the State Guard to be re-created if needed. Maj. Gen. Mark W. Lance,

Florida's adjutant general, planned to fill officer ranks with former Florida National Guard officers, members of the State Guard Reserve, and other qualified persons with "prior suitable military service."[58]

Florida's State Guard Reserve was an inactive component of the Military Department, first authorized in March, 1948. Rather than an organized militia, the State Guard Reserve functioned as a list of former National and State Guard officers who had indicated that they would serve in the State Guard if needed.[59] General Lance stressed to the state legislature and the governor that the state would have to bear the total cost of buying and maintaining uniforms, arms, equipment, and vehicles, if the state chose to recreate the State Guard. For this reason, as well as the presence of some of the National Guard in Florida, the state opted to forgo creating a State Guard during the 1950s.

Florida's reaction to the war in Korea stood on one end of the spectrum of state responses, while Michigan reflected the more aggressive end for states that only began creating a State Guard after the war began. The beginning of the war in Korea led Michigan to revamp its State Troops from World War II.[60] Planning began on August 1, 1950, with a former colonel of the Quarter Master Corps given the state rank of brigadier general and appointed to head the force. Breaking with the World War II standard of creating a State Guard roughly half the size of the peacetime National Guard, Michigan opted to create its Korean War State Guard at one-third of the strength of its National Guard—which meant 8,000 State Troops. This number was actually comparable to the number of State Troops in World War II, reflecting the increases in the size of the National Guard after World War II. The organization of the State Troops called for four Internal Security Regiments scattered throughout the state, with an additional four "special troop companies" based in Lansing.[61]

Records and photos of Michigan National Guard units of the period indicate that it remained a segregated force. Most likely, the Michigan State Troops also reflected this practice, with the four special companies in Lansing being composed of black State Guardsmen. Although President Truman had signed the executive order in 1947 ending segregation in the armed services, actual integration took longer. The active army did not begin to integrate until the added burdens of a segregated army became too cumbersome during the crisis of the early Korean War. The integration of the National Guard took far longer. In many areas, particularly in the South, which had long excluded blacks from joining the

National Guard, blacks were not allowed to join until after the end of the draft in 1973. Since segregation was standard practice in the states that allowed blacks to join, the special troop companies were most likely separate companies of blacks.

Some companies of the two regiments were to assume the state missions of disaster relief and responding to civil disorder that had been assigned to the nondivisional National Guard units based in Detroit and Escanaba that had been inducted into active federal service. The adjutant general, Brig. Gen. George C. Moran, planned to wait until a National Guard unit received its orders for active duty to begin recruiting a State Guard company to replace it. The State Troops so activated had an authorized strength of 1,472 men, although recruiting brought them to only 45 percent of that. The state Quartermaster General issued some arms, equipment, and, unlike in the early years of World War II, trucks. The adjutant general planned to use state funds to purchase uniforms and equipment until the governor received the requested uniforms, arms, equipment, and ammunition, from the secretary of the army.[62]

This Korean War incarnation of the Michigan State Troops never became a fully functioning militia. The adjutant general inactivated the State Troops in June, 1952, after the legislature failed to appropriate funds for it. The disbandment of the force of 733 officers and men came gradually throughout the month of June, with headquarters disbanding last on June 30, 1952. When the force disbanded, the officers and warrant officers received honorable discharges from Michigan State Troops, while the enlisted men were technically discharged from their units. Although the force was never used because much of the National Guard remaining under state control, the adjutant general saw the Korean War experience as a model for future wars.[63]

This was not to be. The State Guard idea, which had been a response to the rise of the National Guard and its dual state and federal standing, was almost extinguished by the Cold War. The State Guard remained for the most part only an idea, usually forgotten, between the end of the Korean War and the mid-1980s. In the age of nuclear weapons, the role of the National Guard in the federal military structure repeatedly came into question. The National Guard needed a period of training lasting a year or more after mobilization before many units would be ready for combat. Strategists believed that in any major war of the future, the United States would need its full force on the first day of the war. The period in

American history when oceans and allies allowed a long period in which a mass army could be readied had ended. In such wars, the National Guard would have little role aside from public order in their home states.

The new emphasis on a large standing army did not lead to the end of the National Guard. Instead, through the lobbying power of the National Guard Association, the National Guard was able to expand its strength and maintain its nominal dual role, but it would seldom be used in its federal war-fighting role; few National Guard units fought in the Vietnam War. The National Guard usually filled the role of home guard itself. The increased employment of the National Guard in domestic disturbances during the 1960s, during racial strife and antiwar protests, fixed for a generation the idea that the role of the National Guard during wartime was in maintaining domestic order. State Guard units that existed after the Korean War performed few missions other than the struggle for survival, and most lost that battle. However, in a handful of states, mostly from the interest of governors and individuals, some State Guard remained in existence, if far removed from the public consciousness, through the years between the Korean War and the 1980s.

A few states maintained an organized militia in addition to their National Guard, usually with the expressed purpose of preparing for the day when the National Guard would again leave the state. Typical of these forces was the Indiana State Guard. Most of the Indiana Army National Guard belonged to the Thirty-eighth Division, which was not activated during the Korean War.[64] But because of the potential that the division might be called, the adjutant created a small State Guard designated as the Indiana Internal Security Corps as a replacement for the National Guard in the event of activation. Under the initial plan, every community with an armory was to have a unit of the Internal Security Corps. In structure the Corps followed that of an infantry division. In its first year, during the early crisis of the Korean War, some 500 officers and 1,200 enlisted men joined. Training, held in two-hour blocks on weekends, at first focused on large-scale civil defense. Later training emphasized traffic control, first aid, guard duty, and limited firearms practice.

Despite its creation as a hedge against a widening of the war in Korea, the Internal Security Corps lasted beyond the war, and longer than the State Guard of most states in the post-World War II era. Permanent permission for states to keep an organized militia in addition to the National Guard came in 1956, with federal legislation, but the new law did little more than legitimize the handful of forces that already existed.[65] Also in 1956, some members of the Indiana Internal Security Corps were activated during floods in the towns of Spencer and Plymouth, on the White and Yellow Rivers, respectively. Despite employment of the force, keeping it alive proved difficult. Year after year the Corps experienced high turnover rates to the point of becoming a revolving door. In his 1955 report, the adjutant general blamed his difficulties in maintaining the Corps on the almost complete lack of support, citing that "it has been extremely difficult to sell the idea of the Internal Security Corps to the citizens of Indiana because of the lack of uniforms, equipment, and pay."[66] The force would remain in limbo for another decade, not quite existing, but never completely ceasing to exist. Only near the end of the Vietnam War would the state again take an interest in its potential replacement for the National Guard.

The Indiana Guard Reserve of the Vietnam era, as the State Guard was redesignated in 1961, was created to be used as a "reservoir of military leaders available for use by the State of Indiana." The force contained a cadre of both officers and enlisted men for expansion in the event of mobilization of the National Guard. On paper the force had three brigades, each with a military police company, and three battalions. Each battalion had five "units." The headquarters company of the Second Brigade operated the armory in Greenfield from the summer of 1968 through October, 1969, while the regular tenant, Company D, 151 Infantry, became one of the few National Guard units to serve in Vietnam.[67] The Guard Reserve, although little known by the people of Indiana, survived through the lean years of the seventies mostly through the efforts of adjutants generals and the few enthusiasts who remained in the force for many years.

Such was the lot for the few State Guard forces that existed between the Korean War and the 1980s. Despite these being the fallow years of the State Guard nationwide, a few states, which had either not created a State Guard during the Korean War or had disbanded one in the early 1950s, created a new State Guard in the 1960s. Often the creation of these forces predated America's increased involvement in Vietnam. The creation of these forces usually resulted from the desire of former National Guardsmen for a continued militia role rather than as a response to the Vietnam War or the unrest of the 1960s. Most of the states that created a State Guard in this period were in the north and largely rural. At the forefront of this small wave of new State Guard creations was Washington, which had created a large State Guard in World War II in response to its fears of a Japanese attack but had created only the barest beginnings of a State Guard at the start of the Korean War. The state on the Pacific Northwest created a State Guard Reserve in 1960 and soon recruited 112 officers, composed mostly of former National Guardsmen, with some business and professional men. The adjutant Maj. Gen. George M. Haskett, created the force to tap into the enthusiasm of the former National Guardsmen, who believed that the possibility of the National Guard leaving the state in the event of a war with the Soviets meant that their services were still needed by their state. The mission of the force was to provide a cadre of officers in case the force needed to be expanded. If expanded because of the loss of the National Guard, the State Guard was expected to work with Civil Defense in civil disturbances, security of public installations, safeguarding emergency food and medical supplies, and providing emergency medical assistance when required, in addition to any of the normal peacetime missions of the National Guard.[68]

The State Guard Reserve was strictly a cadre force, with no missions other than training until ordered to expand. It consisted of a headquarters detachment, two Internal Security Groups, plus the First through Fifth Internal Security Battalions. The battalion was the smallest unit organized while at cadre manning. Each unit contained about a dozen men. These officers trained twice a month at local armories in six communities without pay. They trained in military police duties of arrest, employment of tear gas, marksmanship, as well as general military tasks such as first aid, map reading, and military customs. Twice a year the group held a two-day Command Post Exercise at the state's military reservation, Camp Murray, to prepare themselves to respond to natural disasters or civil disorder.[69]

The force, later renamed the Washington State Guard, with the designation "Reserve" dropped, continued in existence with little change. Unlike most states in the 1960s, Washington was able to maintain its small skeleton force, mostly through the support of sympathetic adjutants generals and a core of enthusiastic members. By the early 1970s, the force maintained 164 officers, who still trained without pay. The basic structure remained the same, although an Air Section with eight men was established at Boeing Field.[70]

For State Guard forces during the Vietnam era, merely surviving became a difficult mission. But with the likelihood of being needed lessening every year, such forces became more of a club of veterans and former National Guardsmen than a real militia force. Providing a role for veterans seems about the only tangible service the State Guard filled in any state. The National Guard did not leave state control, and so State Guard-type forces remained an uncashed and untried insurance policy.

Despite the dismal experience of State Guards in the early 1960s, the war in Vietnam led to some increased interest in militia forces in the second half of the decade. In 1966, the governor of North Dakota began planning for a new State Guard.[71] The strength of the force was not to exceed that of one half of the North Dakota National Guard on June 1, 1940, which placed its maximum strength at 875 men.[72] Despite the governor's intentions, the new State Guard appears to have been stillborn; it was not mentioned in any subsequent reports of the adjutant general. Instead, the reports focus on the Civil Defense agency of the state, and later, the Division of Disaster Emergency Services. These civilian agencies became the vehicles through which North Dakota would react to havoc in the future. Not all states would follow this trend to rely on full-time civilian agencies for traditional militia roles.

Changes in Department of the Army policy toward the National Guard caused at least some states to reconsider their State Guard. Although Nebraska disbanded the last of its State Guard of World War II in 1947, the state statutes for such a force remained valid.[73] The adjutant general annually repeated that the state had the right to conscript men into the State Guard if the need arose. Still, the state maintained no State Guard from 1947 until the mid-1960s. Only during the Vietnam era did the potential to lose the National Guard inspire state leaders to take tentative steps for implementing its long inactive State Guard statutes.

In September, 1965, Secretary of Defense Robert McNamara created the Selected Reserve Force (SRF). The SRF, in theory, contained National Guard and Army Reserve units that were kept at higher readiness levels. These units were to be called into active service ahead of other National Guard and Army Reserve units. Altogether, some 30 percent of the Army National Guard was designated as part of the SRF, which meant less funds for the other 70 percent.[74] McNamara's plan called for these selected units to receive additional funds and first-line equipment, and conduct seventy-two drills that were to last four hours each as opposed to the normal forty-eight drills lasting two hours each. These units were supposed to be kept at 100 percent of their strength in manning and equipment. However, because of the drains on military equipment caused by the war in Vietnam, the Reserve Component units in the SRF never received their full allotment of equipment. By 1967, because of McNamara's belief that the additional training time had not resulted in increased readiness, the SRF plan was scaled back, with the new plan designated as SRF II. Under it, fewer units were involved, but they were supposed to be combat-ready within seven days of a call. Eventually, this reduction of units was followed by the removal of the additional training time, until the whole plan was finally scrapped in 1969.[75]

Despite the shortcomings in the implementation of the plan, National Guard units in the SRF took their role seriously. Nebraska's Sixty-seventh Brigade was part of the SRF.[76] In light of this new role for his National Guard and following authorization from the state legislature, Gov. Norbert T. Tiemann began the re-creation of a State Guard at an initial cadre level in June, 1968, at which time SRF II was in effect.[77] Col. Henry G. Jacoby was chosen to command it. The four-part plan had as Phase I, already implemented, the recruitment of 220 key officers and enlisted men in what was known as the Active State Guard. They served as a cadre for the 35 units in 30 towns.[78] Phase II of the state's plan called for the creation of the Reserve State Guard, containing 1,353 men, which was 35 percent of the authorized wartime strength of the State Guard. At Phase III, as the Standby Reserve, it would have 2,646 men. Finally, at Phase IV, the Augmentation Phase, the State Guard was to be recruited or conscripted to its full strength of almost 4,000 men, including medical teams.

State law called for the force to contain men between the ages of eighteen and fifty-seven, with officers between the ages of twenty-one and sixty-four. Their uniform resembled the army's but with a State Guard shoulder patch, and officers wearing an "N.S.G." brass insignia on their collar. The initial cadre had been recruited in three months from veterans of World War II or the Korean War, most of whom had also served in the National Guard. The state issued them one uniform, a pair of boots, and some organizational equipment. By contrast, officers in the federal military services, as well as the National Guard and reserves, were required to purchase their own uniforms. However, State Guard officers received no pay for most of their training, which consisted of weapons qualification at Camp Ashland—the state's National Guard training area—and training in conducting classes. However, because its reason for being, the SRF, died in 1969, the Nebraska State Guard never moved beyond its initial cadre level. The state disbanded the force in 1972, although the state statutes remained valid.[79]

A similar pattern would be followed in a few other states. A small State Guard would be created as a cadre force, survive for a few years, and wither out of existence. The Maine State Guard of the Vietnam War era stands as one of the larger forces created during that time, despite Maine being a sparsely populated state with few urban concentrations. By the summer of 1969, the force numbered some 97 officers, 6 of whom were chaplains, and 272 enlisted men.[80] The Maine State Guard Reserve (Inactive) included another 11 officers and 5 enlisted men, and twice as many the next year.[81] The force existed as something more than a purely paper force. All officers and communications sergeants, for a total of 95 men, attended Annual School on May 16 and 17, 1970, for communications training. Although this was the only training listed for that year, the adjutant general, Maj. Gen. Edwin W. Heywood, placed enough confidence in the State Guard to declare all units equipped and ready to perform assigned missions, although those missions remained vague.

For the State Guard the legislature appropriated $8,700 annually.[82] However, the force seems to have performed no state service in return for the state's investment. The only decorations handed out recognized attendance rather than service. All members with a year of service received a ribbon.[83] All the men receiving awards were lieutenant colonels or majors, except for one staff sergeant.[84] This gives the impression that the Maine State Guard of the late 1960s was rank-heavy and mission-light. Although nothing written describes it as a cadre force, this conclusion is inescapable. While its members gave up their time to create the force and train together, their actual service to the state seems to have been slight. The Maine State Guard lasted for a few more years but never came into its own as a real force for the state.

By the late 1970s, State Guard, or any state militia besides the National Guard, had largely been forgotten by the general public as well as state and federal governments. Members of state forces served with little or no public recognition or support, training for missions that never came, and with heavy turnover in personnel. Members would explain again and again to family and friends that they were not in the National Guard but another branch of the state militia. The State Guard entered the 1980s as a moribund institution from World War II that had withered but had never completely died.

CHAPTER 10

Total Force and the State Defense Forces

The most recent interest by the federal and state governments in maintaining an organized militia that would remain under state control during wartime began in 1979. The implementation of the Total Force Policy, originally proposed by Secretary of Defense Melvin Laird in 1972, led to an expanded reliance on the Reserve Components—the National Guard and Army Reserve—to wage war. For Gen. Creighton Abrams, the decision during the Vietnam War not to mobilize the National Guard robbed the American war effort of much community-based support. As army chief of staff, Abrams sought in the 1970s to structure the military in such a way as to preclude another Vietnam-level conflict without a call-up of the reserves.[1] Incorporating the Reserve Components more closely into the active army allowed defense planners to maintain a creditable military force in the years following the end of the draft in 1973 and fulfilled Abrams's desire to lock the Reserve Components into participation in any future wars. This placed the National Guard in a position similar to that immediately after the passing of the *National Defense Act of 1916:* in the next war it was to be absorbed completely into the Army and Air Force before any new forces were raised.

The Total Force Policy created a potential need for a state militia in addition to the National Guard. In the decades after World War II, states increasingly relied on civilian agencies, both state and federal, for traditional militia functions. However, state governments still looked to their National Guard to provide immediate service during floods, strikes of state workers, riots, and other emergencies that required a large, disciplined, and flexible organization. As in the world wars, the implementation of the Total Force Policy in any future war would deprive states of their traditional organized militia, the National Guard. Total Force did not introduce a radical concept in American military planning. It simply sought to scale back reliance on a mass standing army and peacetime conscription that had developed after World War II. Unfortunately, the use of the National Guard in mostly domestic functions during the Korean and Vietnam Wars had given Americans the mistaken impression that the National Guard existed primarily for state missions, during wartime as well as in peacetime. Thus the need for, as well as the concept of, a separate state militia without a federal mission had been largely forgotten by the American public, even in those states that had maintained a State Guard in the decades following the Korean War.

The State Guard of World War II and the Korean War had never entirely vanished. Texas, California, New York and a few other states maintained some State Guard-type forces after the end of the Korean War.[2] Not until the early 1980s did these forces begin again to expand in numbers and in visibility. Still, the State Guard-type forces that were planned or actually created remained tiny. Usually manned only at the cadre-level, their projected full-strength levels were also smaller than the National Guard. Because the National Guard existed mainly to fight foreign enemies rather than to perform state missions, it contained far more soldiers than needed for state emergencies. State militia maintained solely for state missions would not have to be as large as the National Guard.

Despite the creation of the Total Force in the 1970s, the new push for State Guards did not begin until the collapse of Detente at the end of the decade. In the late 1970s, some Reserve Component forces began deploying to Europe for their Annual Training, in anticipation of their wartime missions. In 1979, the Soviets invaded Afghanistan, and President Jimmy Carter asked for and received a reinstating of registration for Selective Service, although no draft was undertaken. Not coincidentally, the earliest stirring of a renewed interest in the State Guard occurred in February, 1979, during testimony before the Personnel Subcommittee of the House Armed Services Committee, when a question arose over the status of the National Guard after a mobilization. In response, Brig. Gen. Bruce Jacobs, of Virginia, who was then serving as the Deputy Executive Vice President of the National Guard Association, drafted a memorandum on state troops other than the National Guard. In it he mentioned that little was known about "this most important question" and that much research was needed. The accompanying three-page report noted that thirteen states and Puerto Rico maintained some sort of militia in addition to their National Guard, but no standardization as to organization, training, or even role existed.[3]

Shortly after General Jacobs issued his report, the Historical Evaluation and Research Organization (HERO) began soliciting information from each state military department on State Guard forces from the world wars and Korean War. The purpose was to prepare a larger study on home guards of each state since 1916 under a contract from the Office of the Assistant Secretary of Defense.[4] HERO's hundred-page report, released in March, 1981, made one point abundantly clear: forty-four states had created a militia to replace the National Guard for state service during the world wars.[5] This seemingly obvious point needed to be stressed, because Korea and Vietnam had almost completely erased any understanding of the National Guard's wartime mission from the mind of the American public as well as with most political leaders.

One military leader reading the HERO study was Maj. Gen. George E. Coates, the adjutant general of the state of Washington. General Coates had been appointed by the chief of the NGB to chair a committee on State Guards, and he soon became the NGB's spokesman for its stance on State Guards.[6] At the next conference of the Adjutants General Association, in January, 1982, General Coates briefed his peers on the recent congressional interest in a militia force to fill the state missions of the National Guard after mobilization. He stressed that any third-tier force must not take funding away from the National Guard: any federal money for new state forces must be added on rather than diverted. He also stressed that the State Guard should be manned only at cadre levels before mobilization of the National Guard, in part because he feared that a fully functional force in peacetime would absorb state missions of the National Guard. He called for further study of the missions performed by the State Guard in World War II, as well as of the state missions of the National Guard since the end of the Vietnam War, but he believed a new State Guard should be created to respond to natural disasters and civil disturbances, and to ensure the continuation of vital public services, civil defense, and other specialized missions that might arise after mobilization of the National Guard.[7]

General Coates believed that the federal government needed to take a greater role in equipping these forces. He hoped that obsolete weapons, uniforms, helmets, tactical vehicles, and even helicopters be issued to the states, but that such equipment be maintained in a standby mode by

National Guard technicians. He also called for each state that wanted to maintain such a cadre force to be offered a federally paid technician to plan, organize, and coordinate the State Guard. Under his proposed plan, states would be reimbursed by the federal government for active duty performed by the State Guard, whether for state or federal missions.[8]

By March, 1982, the Department of the Army had drafted its proposed "Policy and Guidance for State Defense Forces," or "SDFs," as State Guard-type forces were increasingly known. Although it was a new document and not simply a rehash of the directives from World War II or the Korean War, it said essentially the same thing. States could maintain such forces if they so desired, the National Guard Bureau would be the Department of Defense's executive agency on all matters pertaining to the State Guard, and membership in an SDF did not make a member ineligible for conscription, but SDFs would not be inducted as units into the federal military. On the uniform issue, the army remained conservative. SDFs could adopt an army-style uniform, but without the distinctive federal insignia and buttons. The name tag on uniforms was to include the full name of the force underneath the wearer's name.[9] However, the regulation had to await legislative action before being adopted, as sections dealt with the issuance of federal property for use by SDFs, and no authority was then in existence.[10]

In the spring of 1982, an article in the journal of the Army War College also called attention to the implications of the federal and state missions of the National Guard. The author, Roger A. Beaumont, a professor of history at Texas A&M, implied that "not all governors are aware of the possible side-effects on their states" of a mobilization of the National Guard, and implied that the National Guard should be relegated to the role of a home guard and wear state uniforms.[11] General Jacobs did not take long in responding to this unwelcome and unpleasant suggestion. In the following issue of *National Guard*, the National Guard Association's magazine and hence the unofficial organ for the National Guard's corporate stance, he pointed out that most governors should be aware of the situation, as the summer 1981, meeting of the National Governors' Association had adopted a statement affirming that the Army National Guard and Air National Guard would fight alongside the active forces in combat. He instead strongly suggested that the National Guard Bureau begin the groundwork for "internal security forces" to take over the National Guard's state missions after mobilization.[12] As throughout the twentieth century, the National Guard sought to protect its state missions, yet at the same time find an institutional solution to fulfilling those state missions during wartime.

The next meeting of the National Governors' Association, in the summer of 1982, saw continued interest in state defense forces. This time, the impetus came from the Federal Emergency Management Agency (FEMA), which had a "major interest in the SDF/SG issue." FEMA had formed a task force under Maj. Gen. John McConnell, a former adjutant general of Indiana, who had been brought out of retirement to lead the effort. FEMA had been directed by the under-secretary of defense for policy to "look at the whole range of matters relating to the formation of State Defense Forces/State Guard in the event of a mobilization of the National Guard." The task force was to lay the groundwork for a coordinating committee FEMA was forming, which included representatives from the NGB, as well as Selective Service, but not representatives from the federal military, an omission that stressed the civilian nature of the FEMA effort.[13] FEMA did not want to be seen as part of the federal military and instead stressed the civil defense and disaster relief nature of its work. The National Guard Association believed that FEMA was "solidly behind the concept" of SDFs, but the NGB later reminded FEMA to work through state adjutants general on military matters rather than attempting to go through the NGB or straight to the state governors.[14]

In September, 1982, at the National Guard Association's General Conference in San Juan, Puerto Rico, the NGA passed a resolution calling for federal support of SDFs. In order to allow SDFs to take over the National Guard's state missions during wartime, the NGA believed such forces needed federal equipment, uniforms, training support, and some "full-time personnel to assist in the day-to-day planning." It noted that the deputy assistant secretary of defense, Reserve Affairs, had been working with the NGB on the SDF issue, and that in the near future, other federal agencies—most likely the FEMA group—would be included in the coordinating committee. The NGA believed it needed to throw its support behind the movement to ensure success, which would help silence those who expected the National Guard itself to perform a home guard role during the next war.[15]

Despite the NGA's desire for federal support for SDFs, the NGA took the stance that an SDF was "solely a state organization." The NGA believed that SDFs could not be brought as units into "the military service of the

United States," and would not be "subject to Federal regulation, control, or supervision."[16] The National Guard Association, which should have known better, had fallen into the same fallacy it had during the world wars. Because the SDFs were militia, the U.S. Constitution gives ample authority for Congress to call them into federal service as a body or as individuals to "execute the laws of the Union, suppress insurrections and repel invasions," and also gave to Congress authority for "organizing, arming, and disciplining the force."[17] SDFs differed from the National Guard because of the dual oath of the National Guard, which allowed it to be part the army and leave the territory of the United States, while no such dual status existed for State Defense Forces. Hence, constitutionally, SDFs could be federalized—brought under the control of the federal government in an emergency—but not drafted en masse into the federal military. Federal laws could specifically exempt the State Guards from federal service, but nothing in the Constitution prohibited federal service for state militia. The attorney general's ruling from 1912 that the National Guard was militia and thus could not leave the territory of the United States now applied to SDFs.[18] However, the National Guard, the army, and the governors were more concerned with state missions than with federal missions, and no challenge was made to the NGA's assertion.

The National Governors' Association continued to be interested in SDFs at their next conference, on December 13, 1982, in Richmond, Virginia. Nolan E. Jones, the staff director of the Committee on Criminal Justice and Public Protection, invited General Jacobs, the deputy executive vice president of the NGA, to speak about the progress of SDFs at the committee's meeting during the conference.[19] General Jacobs had become the National Guard's authority on SDFs, and he described to the committee the progress of the SDFs maintained in more than a dozen states, with more in the planning stages.

Where they existed, modern State Guard units were usually cadre forces containing a mix of veterans, often retirees from the federal military or National Guard, and individuals who had no prior service. Some members had reached an age where they were too old to join the federal military, or did not want to be subject to the global responsibilities of the National Guard, yet wanted to serve their communities in some military capacity. Retired members of the National Guard or federal military provided most of the leadership for the SDFs. The men simply cut off the buttons of their old army uniforms and in most states replaced them with buttons used by the state police, and swapped the black Army name plate for the red SDF name plate.

On the forefront of the renewed interest in the State Guard was South Carolina, which reactivated its State Guard with the active support of the governor, in October, 1981. The new force adopted the organization of a light division of the U.S. Army, being composed of three brigades, each with three battalions. The force had a mix of infantry and military police units, as well as supporting units. That first year saw units formed in 80 communities throughout the state. The initial cadre force contained only 250 men in all ranks, with plans for ten times that number to be recruited, which would allow the force to begin forming companies and even platoons. Leadership came from retired National Guardsmen and Regulars. The state had no uniforms to issue the force, but in the early days the adjutant general believed that federal assistance to State Guards nationwide would soon allow the state to give uniforms to members.[20]

The deputy director of the new South Carolina State Guard was Brig. Gen. John A. Crosscope, Jr., a retired army officer. He had been recommended to the governor by the adjutant general for appointment as a brigadier general in the state military department, with the mission of creating the force. From his position he became immediately involved not only with the creation of a State Guard in South Carolina, but with the national movement for such forces. General Crosscope believed the United States to be under attack and that State Defense Forces were urgently needed. The attack came in the form of hard-core Cuban criminals who had arrived as part of the 1980 Cuban boat lift and from drugs flooding into the country with the active support of Fidel Castro. The Communist world would use every means at its disposal from urban terrorism to thermonuclear war to destroy the United States, and Americans needed to prepare.[21]

General Crosscope also believed that a State Defense Force would perform a role in the All Volunteer Military by providing the military training for young men that had been performed by the Military Training Camps earlier in the century. And here might lie the reason for one of General Crosscope's more unusual assertions. On the seemingly mundane topic of uniforms for the force, he insisted that State Defense Force members have the right to wear the eagle buttons and emblems of the U.S. Army on their uniforms. He based this on his idea that since both the unorganized militia and the organized militia—the National Guard—existed under the same sections of the U.S. Constitution, then both should be able to wear the same uniforms. In the South Carolina State Guard, officers took the same oath of office as did National Guard officers, in which the potential officer swore to uphold the federal as well as state constitution and to obey the president as well as the governor. The official position of the state military department considered the State Guard to be part of the "United States Militia," rather than the organized militia of the state. General Crosscope believed the uniform issue needed to be addressed quickly, for only the federal eagles would attract the higher-quality college graduates on which he hoped to build his force.[22]

The next year, the creation of a national organization of SDF enthusiasts got a boost from Martin Fromm and Associates, Inc., of Missouri, which billed itself as "Association Management Specialists." On November 17, 1983, a meeting of the Steering Committee for a proposed State Defense Force Association was held in Kansas City, Missouri. General Crosscope attended, representing the South Carolina State Guard. Also present were representatives from the Texas State Guard, the Ohio Military Reserve, and the California State Military Reserve. The remaining member of the committee, who did not attend, represented the New York Guard.[23] Together, these five states were the most active in the SDF movement, and all but South Carolina had maintained a State Guard for most of the years since the Korean War.

The committee heard and endorsed General Crosscope's position paper on State Defense Forces. But despite the presence of General Crosscope, the real instigator behind the proposed State Defense Force Association was Jerry Fogel, a vice president of Martin Fromm and Associates. Fogel, forty-seven yeas old, had entered the United States Military Academy in 1954 but did not graduate. He joined the New York National Guard in 1959, and received a commission in field artillery in the summer of 1961. He left the National Guard as a second lieutenant in 1965, and had since worked as an actor and radio personality.[24] For several years Fogel lived in California, where he had a recurring role in the TV show *White Shadow*. In California he became familiar with the State Military Reserve. After his role on the show ended, he entered other fields and started working for Martin Fromm in 1980. Always interested in the military, Fogel became involved with State Defense Forces through his position at Martin Fromm.[25]

That first meeting produced a list of objectives for a future association composed of state chapters of SDF members and supporters. The Steering Committee took great pains to ensure that their proposed organization would in no way be seen as a threat to either the National Guard or the National Guard Association. Fogel assured the committee that he had been in contact with Col. William E. Florence, chief of the Office of Policy and Liaison at the National Guard Bureau, as well as with Maj. Gen. Francis S. Greenlief of the National Guard Association. He had let them know of the plans for an SDF Association and would keep them abreast of all developments. In addition, the committee drew up a list of twelve possible members of a Board of Advisors they hoped to form by the following spring. Included on the list were Sen. Strom Thurmond (R-S.C.), as well as the president of the Adjutants General Association, the president of the NGA, the chief of the NGB, an assistant secretary of defense, the secretary of the army, and the sergeant major of the army.[26]

Despite this ambitious start, cracks were already apparent in Martin Fromm and Associates' attempt to create a State Defense Force Association to act as a national voice for State Guard forces throughout the nation. The Assistant Adjutant General from Washington State had written to Fogel explaining to him that the majority of personnel in the Washington State Guard preferred to be accepted as members in the NGA rather than a new State Defense Force Association.[27] Despite its lukewarm response to the new association, Washington state's military department stood firmly behind the new movement for a State Guard.

Still, when Fogel forwarded the minutes from the meeting to Lt. Gen. Emmett H. Walker, the chief of the National Guard Bureau, he received an encouraging reply. General Walker said that he had much interest in the association and "strongly support[ed] the State Defense Force program." His letter indicated that he believed the association to be a going concern. Fogel made several copies of the general's reply and forwarded them to the members of the steering committee.[28] The next day, December 16, Fogel received an invitation from Col. Florence for the members of the committee to meet with him and General Walker in General Walker's office the following February.[29]

Over the next year, General Crosscope became less and less enthusiastic over Fogel's handling of the State Defense Force Association project. After the November, 1984, meeting of the Steering Committee, General Crosscope broke with Fogel. He objected to Fogel's "sales approach" to creating the association. He believed that Fogel had smoothed out the minutes of the November meeting to give the impression that it had "no jagged edges" and that the association was already functioning. Fogel had also given the impression that the National Guard Bureau had "approved" the State Defense Force Association, which it had not, nor did it have any authority to do so. General Crosscope became convinced that the only beneficiary of the association would be Martin Fromm and Associates, as the management firm overseeing the association. Most members of SDFs already received all the benefits proposed by Fogel through their membership in other organizations, such as the National Guard Association, the Air Force Association, the Association of the United States Army, and the Retired Officers Association.[30] General Crosscope then sent letters to the NGB and the NGA informing them of what he thought of the involvement of Martin Fromm and Associates in the SDF Association, and that South Carolina would not support the State Defense Force Association.[31]

Part of General Crosscope's pique came from Fogel's references to Dr. George J. Stein, a professor of Interdisciplinary Studies at Miami University in Ohio, as the association's consultant, as well as some unauthorized indications that the association had endorsed an article Stein had recently published in *Military Review* that was critical of the SDFs then in place.[32] Stein had suggested that state governments still had not internalized that they would lose their National Guard in the next war and were proceeding slowly in the creation of SDFs. Most likely, General Crosscope was most annoyed by Stein's criticism of the SDFs that did exist. Stein believed that for SDFs to be of any use, they needed to be trained in police functions of riot and traffic control, as well as crisis relocation planning. He believed that basic infantry training for SDFs, such as map-and-compass exercises, stemmed from "nostalgia or fantasy" and was a waste of time. He also drew attention to the potential problems SDFs might encounter when operating in the same area as a federal unit, without any clearly established relationship between the two.[33] General Crosscope called the article "an honest attempt . . . to stress the importance of the State Defense Force program" but said Stein's research was "superficial" and that South Carolina did not endorse Stein's views.[34]

Despite the loss of support from General Crosscope and the South Carolina military department, the State Defense Force Association of the United States (SDFAUS) began functioning the following spring, 1985, with the appointment of the association's first president, and the publication of the first issue of the *SDFAUS News.* States originally involved with the association were Texas, California, New York, Ohio, New Mexico, and Washington, as well as the Commonwealth of Puerto Rico. Although the SDFAUS was based in Missouri, the state had no functioning SDF.[35] Surprisingly, Col. Dean Dubois, chief of staff of the South Carolina State Guard, who had attended the earlier Steering Committee meetings with General Crosscope, was also listed as a board member. Fogel became the executive director of the publication.[36]

The founders created the State Defense Force Association in part to work for State Defense Forces in a manner similar to that which the National Guard Association had worked for the National Guard. However, precious few states maintained an SDF, and not all of those that did were active in the new association.[37] In Fogel's column in that first issue, he stressed that the future of SDFAUS depended on the future policy of the federal government related to missions, uniforms, and funding, as well as the creation of an SDF by every state. Only then, be believed, would the association "be able to exert its full influence upon the national scene." He neglected to mention that the tiny association had almost no control over either issue.[38]

Despite Fogel's desire for federal and state action to make SDFs successful, the movement really rested on local enthusiasm.[39] Average Americans are not attracted to service in an unpaid militia that requires members to buy their own uniforms, especially during peacetime. The early publicity for these new state forces was not always what the army had hoped for, emphasizing the more apocalyptic possibilities of SDF service as opposed to the more likely mission of rescuing people from floods, for which the National Guard was increasingly becoming known. For example, the March, 1985, issue of *Combat Arms* mentioned the new SDFs as an "army of last resort" with missions to "quell domestic disturbances" and "combat terrorism" should the National Guard be sent overseas.[40] A story the next fall in *American Survival Guide* began with a quotation of the "Second Amendment" and proceeded to support SDFs from what it interpreted as a constitutional "demand" that states keep a militia, arguing that the National Guard would not fit the bill during wartime.[41] A cover story in *Military* magazine for February, 1986, was more alarming. It drew parallels between the recent revival of the British Home Guards with a mission to protect their nation against Soviet Spetznaz forces, units of highly trained Communist soldiers intended to perform covert operations in a target area. The writer foresaw American SDFs also fighting Spetznaz teams, as well as terrorists trained in the Middle East or Latin America. Even more ominous was his claim that "thousands of men and women, including Americans, Canadians, and Europeans, have received intensive training in urban guerrilla tactics" in the Soviet Union and were "spread to every continent as instructors."[42] Clearly, for some enthusiasts, SDFs would have their hands full in the next war battling Soviet Special Forces, Latin American Marxists, Middle Eastern religious fanatics, and traitorous Americans.

The 1980s version of the State Guard differed in important aspects from its earlier incarnations. Previous home guard-type forces had been created in response to an actual or imminent loss of the National Guard and the existence of a hot war. Many home guard units had formed before federal or even state authorities began to consider the problem of local defense. But in the 1980s, the creation of SDFs stemmed from federal and state concerns over a theoretical loss of the National Guard. Because SDFs existed to meet potential rather than existing problems, some members were drawn to the organizations more for political and theoretical reasons than for practical ones. The overwhelming majority of members were older veterans of the federal military or National Guard, who brought a maturity and experience desired by all levels in the SDF movement. However, some members were drawn by something quite different.

The mainstream American press also noticed these new militia forces, although coverage remained light and could as often be negative as positive. Occasionally, it could be worse than negative, reaching to condescending and comical. An article entitled "The Few, The Pointless: The Texas State Guard," ridiculed the force as "a third-string military bench warmer" whose primary mission seemed to be granting waivers to allow non-Texans studying at Texas universities to pay in-state rates. Ironically, the Texas State Guard performed some of the few real missions performed by State Guardsmen throughout the 1970s and 1980s, and was one of the least rank-heavy forces. Most SDFs were entirely cadre forces, but Texas had fleshed-out units. Nevertheless, the article criticized it for having one officer for every two enlisted men.[43]

Despite such unwanted publicity, the Texas State Guard was about to have a far worse public relations nightmare that foreshadowed a problem in several other states. The problem stemmed from unresolved issues about the role of the militia in modern America, especially for a contingency force like an SDF. The Texas State Guard included younger enlisted men who had joined for something more exciting than sitting around a classroom during drill or directing traffic at the state fair. The problem of conducting training at a level that kept volunteers interested came to a head in 1984. The issue concerned the 105th Military Police battalion of Fort Worth and, more specifically, its commander, Maj. Robert Holloway.[44]

Major Holloway, then thirty-six years old, had been in the army Special Forces, served two tours in Vietnam, and later worked as a mercenary in Rhodesia. Under his enthusiastic leadership, the sixty members of the battalion began training more hours every month, using their own weapons, including semiautomatic rifles. They earned praise from an Army Reserve unit that had used the 105th as a mock enemy ("Opposing Force" or "Op-For") during exercises. But officials in the State Guard became concerned that the unit was a bit too gung ho. Officials at headquarters saw the wartime missions of the State Guard as being more involved with disaster relief and safeguarding armories than with engaging in combat with invaders or terrorists. Despite recruiting brochures that showed a State Guardsman with an assault rifle and the slogan "Are You Ready for the Task?" peacetime missions were even more mundane than those of wartime.[45] In addition to traffic control at public events, headquarters of the State Guard suggested that National Guard commanders ask State Guardsmen to serve coffee to National Guardsmen returning from training or to wax floors in the armories. The ensuing uproar from State Guardsmen caused the suggestion to be quickly withdrawn, but many State Guardsmen, especially Major Holloway, found the very suggestion belittling and insulting.[46]

The break between Major Holloway and the State Guard came over uniforms. In early 1984, members of the 105th began wearing self-purchased camouflaged uniforms—Battle Dress Uniforms, or "BDUs"—instead of their state-issued olive green uniforms. At the time, Army Regulations prohibited the wearing of BDUs by SDFs. However, Texas state law said that the uniform of Texas state soldiers would be the same as uniforms for U.S military forces.[47] The Texas Army National Guard had switched to BDUs about half a year earlier, so Major Holloway believed he was on firm ground. In the ensuing battle, the Texas State Guard got what it least wanted: lots of negative media attention, with questions raised about the State Guard's budget and sloppy paperwork, and whether such forces were anachronistic. Although some at headquarters wanted to court-martial Major Holloway, a move he welcomed, in the end the State Guard quietly gave him an honorable discharge and considered the problem solved.[48]

The problem was not over for the Texas State Guard, and actually the public relations nightmare had just begun. In October, 1984, a few months after Major Holloway's forced separation, State Guard officials disbanded the 105th, formerly the highest praised unit in the State Guard, for "ineffectiveness and inefficiency," and honorably discharged all its members. Gathering his former State Guardsmen, Holloway formed them into the First Battalion, First Light Infantry Regiment, Texas Reserve Militia, a private group, incorporated as the "National State Defense Force Association," that did not form part of the organized militia of Texas.[49] Holloway spent the next few years either attempting to bring his Texas Reserve Militia back into the State Guard structure or, failing that, persuading the Texas state legislature to stop all funding of the State Guard.[50]

Texas shared with a few other states the difficulties of maintaining an active militia in the modern age and in a nation at peace. The Utah State Guard, reactivated in 1981, had in six years attracted numerous "convicted felons, mental cases, and neo-Nazis" into its ranks, according to Maj. Gen. James Mathews, the adjutant general for Utah. Background checks of potential recruits were almost nonexistent. The discovery that some State Guardsmen also belonged to the Aryan Nations white supremacist group, and at that least one State Guardsmen had traveled to the Aryan Nations' compound at Hayden Lake, Idaho, to train members for combat, demonstrated just how little control the state retained over membership in the force. Unrelated troubles in the cavalry troop brought all the problems of the State Guard into public view.[51]

In a normal army brigade, a cavalry troop operates under the direct control of the brigade commander and is used primarily for scouting the enemy's location. As such, membership in a "cav" troop traditionally carries with it a certain glamour. The commander of the cavalry troop of the Utah State Guard, Maj. Robert A. Kofford, began adding more excitement to the training of his unit. Major Kofford, a veteran who worked in security at Toole Army Depot as his full-time job, had his State Guardsmen practicing for combat, crawling through the woods and rappelling. Many of the men in the unit worked with Major Kofford at Toole, or at the state prison, and began orienting their unit toward fast attack and search-and-destroy missions. Utah's cav troopers began buying and wearing different uniforms from the rest of the State Guard, choosing the older-style solid green jungle fatigues instead of the uniform issued to the rest of the State Guard. The men in the cav troop bought themselves Stetson hats and started wearing yellow ascots, adopting the image of the First Cavalry Division (Air Mobile) as portrayed by the actor Robert Duvall and his men in the movie *Apocalypse Now.* Some of the unit's more eccentric members began referring to themselves as the "governor's own little army," although the governor knew nothing of the force.[52]

After the cavalry troop began conducting its more exciting training, the military police training of the rest of the State Guard suddenly seemed less interesting to members. Other units in the State Guard began holding unauthorized war games in the state's western deserts and practicing assassinations at the fairgrounds. In order to circumvent prohibitions on firearms training, units formed gun clubs with the same members as their State Guard company. State Guardsmen maintained intelligence files on potential "subversives" to be watched if war came. Even more galling to some veterans in the state military hierarchy were the presence of self-identified "elite" forces.[53] One former Army Special Forces soldier in the State Guard hierarchy became incensed when he spotted three members in downtown Salt Lake City wearing camouflaged uniforms, green berets, and insignia similar to that worn by U.S. Army Special Forces. To make matters worse, one of the so-called Green Berets was a woman, and one of the males was "built like a pear."[54]

A far more dangerous incident occurred during the fall of 1985, when a group of men wearing military uniforms and carrying weapons charged from the brush and surprised a group of Girl Scouts on a picnic.[55] By the fall of 1987, General Mathews had had enough problems with the State Guard and disbanded it. He then reorganized it, renaming it the "Utah State Defense Force." The new force contained only about thirty officers, much fewer than the four hundred or so men the State Guard had in the mid-1980s. The new SDF had as its only mission the safeguarding of National Guard armories should the National Guard be called away. Rather than conducting attacks against the enemies of the state, the new SDF would be checking locks and turning off the lights at the local armory if the National Guard left the state. In this reduced form, General Mathews believed the Utah State Defense Force was "functioning without difficulty and [was] providing significant assistance in the areas of mobilization and emergency support" during the partial mobilization of the National Guard in the fall of 1990.[56]

In another age, or in the federal military, officers like Holloway and Kofford most likely would have been praised for their imaginative and active leadership. They had instilled a fierce sense of unit loyalty among their men, and developed exciting and challenging training at little or no cost to the state. But the 1980s were not 1775 or the early years of World War II. Men training for war, who were not part of the federal military, appeared either comical or dangerous to much of the American public. But men like Holloway and Kofford were far from alone in their vision of SDFs fighting subversives on the streets of America.

At the fourth annual conference of the SDFAUS on October 29, 1988, in Kansas City, Missouri, SDF members listened to presentations by an officer of the Defense Intelligence Agency on "Soviet SPETZNAZ and Protection of Key Assets," "Reflections on International Terrorism," and "Perceptions of the Soviet Military."[57] Stein, whose dismissal of the combat potential of SDFs in his 1984 article had earlier cost the SDFAUS the support of General Crosscope, had changed his position considerably over the next four years. In his keynote speech at the 1988 conference, he attacked America's "political elites" and its easily led public, who had built all of American defenses on the idea that "either strategic nuclear deterrence works, or the world ends." As a result, he believed, America had left itself vulnerable to the threat of Soviet saboteurs and Spetznaz attacks. The U.S. military was geared totally to respond to threats at the national level, when in reality, he claimed, the attacks would be local, that is, small attacks inside the United States. He claimed the national military "neither structurally, *nor legally,*" [emphasis his] had a domestic role. But the public had "no consensus that there [was] any domestic threat for which SDF is needed," and worse yet, America had an "anti-SDF tradition." He urged the SDFAUS members "to force the relevant elite decision-makers at both the national and state levels to *even recognize* [emphasis his] that (a) there is a continuing threat; (b) there are domestic targets; and (c) our enemies increasingly have the means to attack these targets."[58] Only then, he believed, would state and federal support of SDFs be possible.[59] In the four years since his "Missing Link" article had been published, Stein had come to General Crosscope's position on SDFs. The American public, as well as state and federal military leaders, would soon move in the opposite direction.

The 1984 movie *Red Dawn* had presented a story of an America in the near future in which Communist forces from Latin America and the Soviet Union invade and occupy large sections of the interior. The movie follows a group of high-school-aged friends as they wage guerrilla warfare against the occupying forces. The movie was immensely popular among Americans who longed for a break from the dullness of a society long at peace. Some young and not-so-young Americans prepared for the day when they too would have to fight against Soviet invaders.[60]

The collapse of the Warsaw Pact in 1989 and 1990, and the breakup of the Soviet Union that followed, did nothing to stem this yearning. People who needed an enemy looked at the world, searching for a new bogeyman to replace the Soviet Union, and their gaze fell upon the United Nations or, increasingly, the United States government itself. Out of a lingering and often poorly understood conception of a state militia as a bulwark against federal tyranny, such people were originally attracted to the SDFs. Later, as they discovered the more mundane missions for SDFs that would be tolerated by state military officials, they tended to drift away. Despite the cutbacks in projected missions for SDFs, incidents, however minor—like when a member of the newly formed Nevada National Guard Reserve, on his own initiative, wrote a letter to a survivalist magazine soliciting new recruits—became ammunition for opponents of militia in any form besides the National Guard.[61]

One of those opponents was Ed Connolly, who wrote a series of articles on the California State Military Reserve in the late 1980s for the *San Francisco Bay Guardian.* Connolly had become one of the leading spokesman against the whole SDF movement and would later testify before the California Assembly during an unsuccessful attempt to end funding of the force.[62] Connolly took great pains to show the difficulties with the White Supremacists in the Utah State Guard and the problems with Holloway in Texas. He also drew attention to the links between these groups and publications such as *Soldier of Fortune,* which purports to be a magazine for mercenaries. In addition to a nine-page article *Soldier of Fortune* published on Holloway, Connolly found that a visiting editor from the magazine had been made an "honorary major" in a California unit after he participated in unauthorized war games in the national forests of California. But the rogue elements within the State Guard were only half of his worries. Connolly also feared the Military Reserve's wartime role in maintaining order. Particularly worrisome were provisions in state law that allowed California to draft men into the force. For Connolly, SDFs originated in the first Reagan administration, when the Department of Defense "courted state legislatures" to create such forces. SDFs became, for Connolly, a cadre force that during wartime would draft young men, train them in the use of truncheons, batons, and M-14s, and use them to crush opposition to the federal government's military policies.[63]

After the end of the Cold War, interest in SDFs declined at the state and national level. After the creation of the Michigan Emergency Volunteers in 1988, no new SDFs would be created. SDFs survived because of the enthusiasm and dedication of members, rather than from federal or state support. Most of the more apocalyptically focused members had been removed, and SDFs settled into an existence more in keeping with their original concept, namely a force of mostly retired Regulars and National Guardsmen who maintained a cadre militia as a hedge against loss of the National Guard.

The mobilization of part of the National Guard for Operations Desert Shield and Desert Storm (the war against Iraq) placed few new burdens on SDFs. Existing Regular and Reserve Component forces fought the war, and no new forces were raised. Most of the National Guard remained in the United States, and all states retained control over enough of their National Guard that SDFs had few additional missions, although some did take on new tasks. For example, during 1990 and 1991, the Georgia State Guard, with just under nine hundred members, suspended all training and shifted to support for mobilized National Guard units. In addition to providing administrative help for Family Assistance operations, the State Guard also furnished the caretakers for vacated armories.[64] Members of the Maryland Defense Force provided Family Assistance for families of National Guardsmen in the 290th Military Police Company after that unit mobilized for the war.[65] Many SDF members had long experience with the military and were well suited to help the families of National Guardsmen deal with the federal bureaucracy that had been designed to assist them during mobilizations, from filling out complicated forms to shopping at the local commissary. But the small war never fully tested Total Force, leaving many Reserve Component units unmobilized, and so it never fully tested SDFs.

After the National Guard units that had mobilized during the war with Iraq returned from active federal service, SDFs settled into a less ambitious posture. An article in the spring of 1991 in *American Survival Guide* by Capt. Allen A. Moff of the Ohio Military Reserve underscored the point. Despite the article's publication in the type of publication higher-ranking officials in SDFs dreaded, Captain Moff's article could hardly have appealed to undesirable potential recruits. Captain Moff emphasized links between the SDF movement and its earlier incarnations in the world wars rather than harking back to some militia of the Revolutionary era or Anglo-Saxon England. Instead of emphasizing the Second Amendment, Captain Moff discussed the unpaid nature of the service and how it had answered a need from within the thirty-seven-year-old nonveteran to perform some form of military service. Ironically, Captain Moff had first heard of SDFs from Larimer's letter on the Nevada Defense Force in the May, 1987, issue of *American Survival Guide.*[66]

Despite the toning down of projected roles for SDFs, old problems would continue to haunt the movement. Connolly would write other articles on other SDFs, attacking in turn the Ohio Military Reserve, the New York Guard, the Massachusetts State Guard, South Carolina State Guard, and others.[67] In all cases, Connolly focused most on efforts at intelligence gathering in SDFs. He feared the SDFs to be a holdover from the Reagan years, when Attorney General Ed Meese and the Department of Defense quietly began to put pressure on state governments to create forces that would one day crush dissent on the streets of America.[68] Columnist Jack Anderson also took aim at these new state militias. Like Connolly, Anderson saw SDFs as unwelcome holdovers from the Reagan administration and the late Cold War. He believed their original mission to have been, in part, "to play a role in FEMA's contingency plans to keep the government operating during a nuclear war."[69] Worse than that, these groups, which he believed had "attracted some Rambo wannabees, neo-Nazis, and frustrated weekend warriors," still had supporters in Congress, specifically, Rep. Floyd Spence (R-S.C.) and Senator Thurmond, who had backed federal legislation to allow these "forces without a function" to use federal military training facilities and to be armed with leftover federal weapons. As with almost all critics, Anderson always included in his criticisms the same list of problems with overzealous SDF members and ties to right-wing groups in Utah, even though most of those problems had been addressed.[70] Still, problems with some members of SDFs in the 1980s would continue to haunt such forces into the 1990s.

A series of articles in Ohio newspapers on the Ohio Military Reserve (OMR) exposed some of the earlier difficulties state military officials had with overzealous members of the force. Sometime in the late 1980s, in an incident reminiscent of the Girl Scout fiasco in Utah, "renegade" members of the OMR, complete with automatic weapons firing blanks, painted faces, and wearing their OMR uniforms, crashed from the woods in a state park near some startled campers. The article by Captain Moff in *American Survival Guide* was also cited as an example of the lack of control the state had over its militia. The newspaper stories did explain that the OMR had abandoned its original counter-terrorism plans in favor of a military police role but still managed to quote Connolly on what he thought of SDFs in general and to mention that state law gave the governor the ability to draft men into the force.[71]

A follow-up story the next day explained the moves by Representative Spence and Senator Thurmond to acquire more federal support for such forces. In this endeavor they were assisted by Rep. Walter B. Jones (D-N.C.), who based his support on the large numbers of National Guardsmen that left his state during the war against Iraq. He believed that a properly trained and equipped SDF could have done more for the state during the National Guard's absence, although he did not specify what additional services the SDF could have provided. Senator Thurmond introduced the bill by drawing attention to the downsizing of the U.S. military, implying that with fewer active forces, the Reserve Components would play an even larger role in the next war. For Senator Thurmond, the period after the war with Iraq was a time to increase rather than shrink SDFs.[72] And for a few states, this was true. The more controversial training, missions, and members had largely been purged from SDFs. Some SDFs, like the Tennessee Defense Force, had little or no trouble in their own states and looked forward to a quiet future of slow expansion and additional missions.[73] At the association, the end of the Cold War had also brought new direction.

In the late fall of 1992, the SDFAUS severed its relationship with Martin Fromm and Associates and moved from Missouri, where it had been based since its inception, to Maryland. Previously Martin Fromm and Associates had been working with the SDFAUS for only a small management fee. New management at Martin Fromm dropped all clients that could not pay normal fees.[74] The SDFAUS did not generate enough capital to become a paying client of Martin Fromm.

The new executive director of the SDFAUS was Col. Paul T. McHenry, Jr., of the Maryland Defense Force.[75] Under Colonel McHenry, the SDFAUS and its quarterly journal became reinvigorated. During this period, Marylanders had the most influence in the association. In addition to Colonel McHenry, the new association president belonged to the Maryland Defense Force, as did Col. Robert M. Gertz, who in civilian life belonged to the faculty at the University of Maryland, and in the association headed the Editorial Committee. Under Colonel McHenry's leadership, the *SDFAUS News* became the *Militia Journal.* He tried at the same time to get the membership to change the name of the association to the "State Militia Association of America," which he believed better explained what the association was about.[76] Although the association eventually rejected this name, it did later accept the "State Guard Association of the United States" (SGAUS), which had the advantage of sounding familiar and the disadvantage of confusion with the National Guard Association.[77]

Under the leadership of Colonel McHenry and Colonel Gertz, the association functioned smoothly and its publication took on a more professional quality. Each issue featured stories from state chapters on recent events. From this source, leaders in each state could see how other states employed or failed to employ SDFs. Most of the stories from the states concerned new training programs or participation in ceremonies, but occasionally an SDF reported actual employment in an emergency, as on February 21, 1993, when thirty-five members of the 402d battalion, 3d Infantry Brigade of the Tennessee Defense Force, assisted in relief operations for the town of Lenoir City, which had been struck by a series of tornados.[78] Still, Colonel McHenry and Colonel Gertz wrote large amounts of each issue.

Despite the prominence of Marylanders in the association, the Puerto Rico State Guard was the largest and most active of all the SDFs. As in World War II, the Puerto Rico State Guard (PRSG) ran ahead of the rest of the nation in innovation and enthusiasm. The 1992 and 1998 conventions were held on the island. With more than 1,500 members, the PRSG was able to perform peacetime support for the National Guard, providing medical examinations and legal assistance for National Guardsmen and serving as instructors at the island's Officers Candidate Academy.[79] Between 1994 and 1999, Puerto Rico provided three presidents for the association. In addition, their Friday night receptions at each convention became a highly anticipated tradition.

The annual conventions showcased the moral support enjoyed by the State Defense Forces from the federal government. At the 1993 convention in Baltimore, the three contenders to be the next chief of the National

Guard Bureau were featured speakers.[80] At the next convention in Austin, Texas, D. Anne Martin, the deputy director, State and Local Preparedness Division of FEMA, spoke to the association about the need for interstate compacts to allow state forces to assist each other in emergencies.[81] The State Guard Association, as well as SDFs in the approximately twenty-five states that maintained them, had weathered the worst of their problems with overzealous personnel.[82] State governments were becoming less interested in SDFs, but the future held new public acceptance and, possibly, responsibilities.

Then the Alfred P. Murrah Federal Building in Oklahoma City blew up on April 19, 1995, on the second anniversary of the destruction of the Branch Davidian compound in Waco, Texas. In the first days after the bombing in Oklahoma, congressmen and the media, as well as most Americans, suspected the attack had roots in the Middle East. However, the date of the attack aroused suspicions, and soon web sites and e-mails on the internet filled with messages from members of private, unpaid paramilitary groups, calling themselves "militia," who interpreted the bombing as the start of a war against federal tyranny. In the media frenzy that followed, these groups received far more publicity than ever before. Links between the two main suspects in the bombing, Timothy McVeigh and Terry Nichols, to the so-called Michigan Militia threw such groups on the defensive. These paramilitary groups began to develop a new theory far removed from their earlier interpretation of the bombing. Rather than see the bombing as the vanguard of a great uprising against the government, the bombing became part of a government conspiracy to build public support for a campaign against these "militia" groups.

Most important for the SDFs, public misunderstanding of the term "militia" placed SDF members on the defensive. Members who had long taken pride in telling their neighbors and coworkers that they belonged to the state militia now found those same neighbors and coworkers looking at them suspiciously. Colonel Gertz removed the word "Militia" from the name of the *Militia Journal* because of the word's negative connotations.[83] Something so dear to SDF members, the title of militia, now carried with it a subversive and dangerous image.

Even more damning for SDFs were the earlier links between at least some of the rogue militiamen and the SDFs. Although SDFs had long since purged the few members who saw state militia as a means to oppose the federal government, the earlier connections would continue to taint the very term "militia," and the SDFs as well. The SGAUS quietly changed the name of its publication from the *Militia Journal* to the *Journal of the State Guard Association* in the summer of 1995, and again to the *SGAUS Journal* the following spring. Colonel McHenry, a lawyer by profession, took great umbrage at what he labeled the "semantic infiltration" of the term "militia."[84]

The problem came from both the antigovernment or, as they called themselves, "patriot" groups as well as their opponents. The so-called militia groups defined themselves by the Second Amendment, which seems to equate the terms "militia" and "the people." These groups tended to ignore the other sections of the Constitution related to militia, instead drawing their "legitimacy" from the *Militia Act of 1903* and state laws that defined the Organized Militia as people in the National Guard, Naval Militia, or State Defense Forces, and declared all other adult males not in the federal military to compose the Unorganized Militia. On the opposite side, Morris Dees, of the Southern Poverty Law Center, headed the "Militia Task Force" to keep an eye on the estimated two hundred such groups in the United States. With the mainstream media also referring to any gang of rural whites with guns as "militia," the term was lost to men like Colonel McHenry, who had championed the concept for so long.

The SDFs were in a no-win situation: they drew their legal and historical right to exist from their status as militia, yet the term had become a dead skunk around their neck. For Colonel McHenry, part of the possible solution to the presence of these so-called militia groups was for state military departments to bring those groups that were genuinely concerned with law and defending the nation into the SDF structure. This certainly had precedence in World War II when similar groups were absorbed as the State Guard Reserve.[85] But if the SGAUS had to struggle to present itself as the spokesman for legitimate militia organizations, the situation for SDFs in many states was worse. The same handful of stories about unauthorized training in national forests, ties between members and White Supremacist groups, and a unit in Virginia that had pooled its funds to buy some sort of armored vehicle, had taken new weight after Oklahoma City. Too many Americans could not differentiate between legitimate state militia and private unpaid paramilitary groups. State governments seemed more anxious to rid themselves of many trained, mature, and eager volunteers rather than suffer the possibilities of giving any hint of sanction to a handful of loose cannons.

The California Military Reserve was particularly vulnerable. When it tried shortly after the Oklahoma City explosion to gain authorization to raise private donations and assist the National Guard in civil disturbances, even the California National Guard argued against it. Newspaper stories erroneously reported that the force had never been mobilized for an emergency in more than fifty-five years of existence, forgetting its active service to the state in World War II.[86] Incidents such as members showing up armed and uniformed at the L.A. riots without invitation, recruiting advertisements in *American Survival*, and one member who had tried to direct traffic from his personally owned jeep during the Oakland fires, presented an image of a force somewhere between comical and dangerous.[87] But many of the incidents showed a natural desire of members to perform some mission, any mission, for their state. Most people joined SDFs because they wanted to be called on to serve, not to train for missions that never came. SDFs that survived this period had to downplay to the point of absurdity the very idea of ever carrying arms. Instead of maintaining public order, SDFs began increasingly to focus on providing communications in their states during emergencies and on supporting National Guard units during deployments by assisting family members and providing caretakers for armories.

Despite such difficulties, activities of some SDFs continued throughout the 1990s. In May, 1995, eleven members of a forty-six-person group designated by the Department of Defense to represent the United States in Poland at the fiftieth anniversary of V-E Day belonged to SDFs. The delegates, from South Carolina and Maryland, wore their state uniforms at the ceremonies.[88] By 1995, the association contained twenty-one state chapters and was actively courting SDFs from four states that did not belong to the association.[89] And in some states, the legislature took action to prevent any unauthorized group from taking the name "militia." Rhode Island, whose Chartered and Non-chartered Commands had returned to the records of the adjutant general, passed a law to protect the name "militia" from any more discredit within that state.[90]

Ironically, despite the end of the Cold War, the potential for a full mobilization of the National Guard remained as certain, or even more so, as during the 1980s. The vast cutbacks in the size of the federal military following the war with Iraq, coupled with the Pentagon's insistence on maintaining the ability to fight two nearly simultaneous regional conflicts, meant as much dependence as ever on the Reserve Components.

By the late 1990s, the State Guard Association was made up mostly of veterans, and retired soldiers, from both the Regular Army and the National Guard. As such, the association was hardly a bastion of opposition to the National Guard or standing army. Instead, the association functioned both as a pronational defense group as well as a lobby for state defense forces. The State Guard Association of the United States held its annual convention for 1995 in October at the Marriot Hotel in North Charleston, South Carolina. On Saturday night, October 21, the association held its annual banquet. Most of the attendees wore their dress blue uniforms or military tuxedos for the event, with the red name tags of SDFs and state buttons instead of the federal eagles. The decorations worn by these members of various State Defense Forces attested to the long years of federal service of many. At one point the featured speaker, Lt. Gen. Edward Baca, chief of the National Guard Bureau, commented on the numerous decorations he had seen and asked veterans of various wars to stand and be recognized. Eventually, almost all of the generals, colonels, and sergeant majors were standing. Therein rests one of the attractions and functions of the modern SDFs and the association: to give these old soldiers an opportunity for camaraderie and wearing their uniforms, while at the same time continuing to serve. That these old soldiers might wear their uniforms at banquets and ceremonies does not bother the National Guard or the army, but the desire of State Guardsmen to perform real missions runs into increasing opposition.

The question remains as to whether they would be able to serve at all in the future. The movement to create new SDFs in the early 1980s had the support of many state adjutants general. Their support stemmed from a variety of reasons, from fear of losing their entire force in a war to fear of federal monopoly over military forces to simply sensing what the Pentagon wanted. After the end of the Cold War the Pentagon lost interest in SDFs. Adjutants generals of the 1990s were less supportive of the SDFs than had been the adjutants general who created them in the 1980s.

With the army indifferent, and the National Guard and adjutants generals indifferent or hostile, SDFs increasingly have had to appeal to the governor or to friends in the state legislature to remain alive. Although an awkward situation, it does mirror the former tendency of the National Guard to bypass the army and appeal directly to Congress. Adjutants general in some states, unable to abolish their SDFs, have "reorganized" them to the point of practical dismemberment. The adjutant general of North

Carolina told the force in April, 1996, not to meet again until he directed it to do so, while the Michigan Emergency Volunteers was reorganized to include only ten members. The Maryland Defense Force had to appeal over the head of its adjutant general, Maj. Gen. James F. Fretterd, to Governor Parris Glendening to be allowed to continue in existence.[91]

When General Fretterd had addressed the association back in 1993 when it met in Annapolis, Maryland, he indicated a long future for SDFs working in partnership with the National Guard.[92] However, in later years he took away the Maryland Defense Force's funding and tried to ban them from furnishing crowd control or traffic services. Eventually the state attorney general overruled most of the legal basis General Fretterd had cited.[93] Maryland's militia may continue to exist, but its future remained precarious without the support of the adjutant general, who is, after all, the governor's main advisor in military matters and the head of all state military forces.

More troubling at the national level, the old prejudices had not completely gone away. The SGAUS was pleased when representatives from the television show *Dateline NBC* contacted several SDFs and planned to come to the fall 1996 convention in San Francisco.[94] Reporters visited Virginia, Utah, and a few other states, interviewed members, and took video of SDF members training. Reporters came to San Francisco and taped the opening ceremonies and conducted more interviews.[95] Association members eagerly awaited the NBC story, which they hoped would clarify to America who they were. The eventual story, which aired on June 17, 1997, could hardly have been worse.

NBC news charged that the SDF program, which it quoted Representative Charles Schumer (D-N.Y.) calling an "ill-conceived program," spawned the so-called militias. The story also featured Lawrence Corb, who had been an assistant secretary of defense in the early years of Reagan's presidency. In explaining his original concept of the SDFs, Corb mentioned that they were to be like the home guards of the world wars who "were block monitors, who would go around and tell people to put the lights out in case there was an attack." Corb, whose office had contracted for the original HERO study, must either have never read the study or forgotten its contents. Despite footage of air raid wardens NBC aired with the segment, neither home guards nor State Guards were synonymous with air raid wardens during the world wars. These earlier incarnations had been armed and had missions far more aggressive than those of the SDFs in the 1980s and 1990s.[96] Yet NBC news gave the impression that modern SDFs were part of Reagan-era paranoia over communism, and far broader and more militaristic than anything created in the past.

Dateline went on to list the same old difficulties with the battalion in Texas and the links to white supremacists in Utah. From the *Dateline* story, SDFs appeared to be highly trained operatives, who undertook Ranger and sniper training. The story claimed that former SDF members then took their training, which NBC News indicated had been conducted by the U.S. government, and joined rogue militia groups. The episode showed none of the footage taken at San Francisco, showing instead clips of National Guardsmen marching, and members of the U.S. Army's ceremonial unit in D.C. It also showed details like an empty office that served as headquarters of the inactivated Utah State Defense Force. Perhaps as a sop to the men at the SGAUS who had so readily talked to the reporters, the story ended by mentioning that not all states with large rogue militia groups had SDFs and that "many volunteers tell *Dateline* they're dedicated to serving their states and their fellow Americans and anxious to be disassociated from what they call rogue militias."[97]

The SGAUS expected fallout after the *Dateline* episode aired, but the lack of any response reflected the widespread public apathy toward SDFs and militia in the late 1990s. Despite their increased marginalization, SDFs continue to have their supporters in Congress and in state houses. In modern America, militia can only thrive with support from federal and state governments. Militia by definition is an armed force, yet modern America seems uncomfortable with employing part-time military forces in maintaining domestic order. An unarmed militia can hardly be called a militia at all, but disaster relief increasingly becomes the only mission available to these forces. The National Guard remains hostile to any military force that would absorb state missions during peacetime, and so consistently blocks attempts by SDFs for real-world missions.

For survival in the future, militia might again have to look to local government for legitimization. Colonel McHenry sees a possible future of militia in locally based self-armed home guard units that could, for example, protect businesses during a breakdown of services caused by a coordinated sabotage of power systems. Such local forces might be able to respond faster than the National Guard. Colonel McHenry rejects the idea of gearing militia toward search-and-rescue-type missions. He believes that any group of civilians can perform search-and-rescue, but only militia, which by definition is armed, can protect property.[98]

At the end of 2000, Colonel McHenry stepped down as executive director of the SGAUS. The last decade had left him pessimistic over the future of SDFs. No new states had created viable SDFs since 1989, and Michigan, Nevada, North Carolina, and other states either reorganized their SDF until it no longer existed or disbanded it completely. Although the SGAUS would continue to exist, and Colonel McHenry would remain a member, the headquarters of the association would travel with the annually elected president. The annual meeting and annual newsletter had become the only activities of the association.[99]

SDFs in the nineteen or so states where they continue to exist draw a dedicated mixture of unpaid volunteers with prior military service, and others who had not served in the federal military and are looking for a taste of what they missed.[100] The SDFs were created to provide a state military force for those periods when the National Guard is fulfilling the other half of its dual role. And in the era of a shrinking U.S. military, they continue to prepare for the day when the National Guard again leaves the states for distant battlefields, while attempting to develop new reasons for existing in the event that the entire National Guard never leaves. As such, they represent the latest chapter in a long struggle over the proper role for militia in the United States.

Conclusion

The dual oath or enlistment that National Guardsmen have taken since 1916 allows the National Guard to accompany the Regular Army and Air Force to any theater in the world. Although states continue to have some measure of control over their National Guard, the amount continually decreased throughout the twentieth century as the federal government paid an ever greater share of its support. The defeat in federal courts of an attempt in 1987–88 by the governors of Minnesota, Massachusetts, and other states to prevent their National Guard units from attending annual training exercises in Honduras underscored the point.[1] Governors of those states had attempted to resist what they saw as the Reagan administration's use of the National Guard to pressure the Sandinista government of Nicaragua by demonstrating the ability of the United States to put large numbers of American soldiers on its border. The governors' defeat made clear that the National Guard is a federal force that is available to the states only at the pleasure of the federal government. Governors hold command over their states' National Guard only when the federal government permits it. Federal missions have an absolute priority over state missions, in war or peace. State governments become little more than cheerleaders for their National Guard units once they enter federal service.

States have lost their entire National Guard to its federal mission twice in the twentieth century, from 1917 through 1921, and again from late 1940 through 1947. The states also lost large numbers of their National Guard during the Korean War, the Berlin Crisis of 1961, and the war with Iraq in 1990–91. Although the likelihood of the federal government again calling the entire National Guard into federal service is unlikely, the

National Guard is increasingly geared toward its federal war fighting role. As a result, the National Guard spends little time preparing for its state missions. Although an unarmed militia may be an oxymoron, purely state militia can be geared toward those local missions that are increasingly overlooked by the National Guard. The State Defense Forces were but the latest attempt to create a state militia that will serve locally and not be lost to the federal war effort during wartime.

The American militia tradition holds that all able-bodied free adult men have an obligation to bear arms when called to do so by their government. While governments are hesitant to call on any militia besides the National Guard, modern State Guardsmen are eager to serve. Whether they will continue to have a role in the twenty-first century depends, as always, on their ability to maintain competent forces without being a drain on the federal military or a threat to the National Guard.

Notes

Introduction

1. See the cover story in *USA Weekend*, the magazine supplement to the *Sunday Times (USA Today)*, for Apr. 12–14, 1996, for an example of this.
2. John C. Fogerty, "Travellin' Band," Jandora Music.
3. The National Guard's unofficial historian, Jim Dan Hill, made no mention of militia forces raised for the home front during wartime in the twentieth century in his *The Minute Man in Peace and War: A History of the National Guard* (Harrisburg: Stackpole Company, 1964). The standard work on the American militia and National Guard, John K. Mahon's *History of the Militia and National Guard* (New York: Macmillan, 1983), contains only a few fleeting references to home guard-type forces. The authoritative bibliography on the American militia, Jerry Cooper's *The Militia and National Guard in America since Colonial Times: A Research Guide* (Westport, Conn.: Greenwood Press, 1993), notes only one article covering such a force.
4. The most prolific writer of such articles is Merle T. Cole, a former officer in the Maryland State Guard, who published several articles on militia forces in West Virginia and Maryland during the world wars. See "Martial Law in West Virginia and Major Davis as `Emperor of Tug River,'" *West Virginia History* 43 (Winter, 1982): 118–44; "The Department of Special Duty Police, 1917–1918," *West Virginia History* 44 (Summer, 1983): 321–33; "Organizational Development of the West Virginia State Guard," *West Virginia History* 46 (1985–86): 73–88; "Maryland's State Defense Force," *Military Collector and Historian* 39 (Winter, 1987): 152–57; "The Special Military Police at Morgantown Bridge, 1942–1943," *Military Affairs* 52 (Jan., 1988): 18–22; "Second Regiment, Infantry, Maryland State Guard: An Early State Defense Force," *Journal of Military History* 53 (Jan., 1989): 23–32. See also, Charles R. Fisher, "The Maryland State Guard in World War II," *Military Affairs* 47 (Jan., 1983): 11–14; Christopher C. Lovett, "Don't You Know There's A War On? A History of the Kansas State Guard in World War II," *Kansas History* 8 (Winter, 1985–86): 226–35; Raymond Johnson and Richard P. Ugino, "21st Regiment, New York Guard, 1940–1945," *Military Collector and Historian* (Winter, 1983): 178–79; and Margaret Miller's hatchet job, "Yankee Doodle Dandies . . . or Doodle Duds?: The California Militia Awaits the Call," *California Journal* (Aug., 1992): 379–84.

   For more information on the militia in Texas, Puerto Rico, Indiana, and Michigan, see Luis A. de Casenave, *Historia De La Guardia Estatal De Puerto*

Rico (1508–1992): Resena Historica del Origen de La Guardia Estatal de Puerto Rico (San Juan: Guardia Estatal de Puerto Rico, 1992); Valentine J. Belfiglio, *Honor, Duty, Pride: A History of the Texas State Guard* (Austin: Eakin Press, 1995); Felix E. Goodson, *The Indiana Guard Reserve* (Indianapolis: Indiana Creative Arts, 1998); and, Duane Ernest Miller, *Michigan's State Defense Forces: A Concise History and Lineage of the Michigan Emergency Volunteers and Its Predecessors, 1917–1998* (Lansing, Mich.: Berlin Green Press, 1999). The above authors were members of their respective State Guards.

5. See Curren R. McLane, *A History of the Texas State Guard* (Austin: Headquarters, Texas State Guard, n.d.); *History of the California State Guard* (Sacramento: State of California, The Adjutant General, 1946); and, *History Massachusetts State Guard,* a manuscript apparently self-published by the Massachusetts Military Department in 1978.
6. See Stephen D. Johnson, and Gary S. Poppleton, *Cloth Insignia of the U.S. State Guards and State Defense Forces* (Hendersonville, Tenn.: Richard W. Smith, 1993).
7. George A. Craig, *History: I Massachusetts State Guard, II Massachusetts State Guard Veterans* (n.p.: George A. Craig, 1931).
8. The only full-length work on militia forces raised during wartime in the twentieth century is the study by the historical think tank HERO, *U.S. Home Defense Forces Study* (Dunn Loring, Va.: Historical Evaluation & Research Organization, Prepared under contract for the Office of the Assistant Secretary of Defense, 1981).
9. The latest article dealing with this apparent paradox is George J. Stein's "State Defense Forces: The Missing Link in National Security," *Military Review* 64 (Sept., 1984): 2–16.

Chapter 1

1. The English militia can be traced to the Saxon Fyrd, which enrolled free men for training in the use of weapons for local and national defense. Over the years, the Fyrd developed into the General Levy, which, like the Fyrd, was based on the obligation of all able-bodied free men to perform military service within their own districts and, in the event of invasion or rebellion, march to any part of the kingdom for service. See C. Warren Hollister, *Anglo-Saxon Military Institutions on the Eve of the Norman Conquest* (Oxford, England: Oxford University Press, 1962).
2. For more information the use of the militia to enforce revolutionary discipline during the War for Independence, see John W. Shy, "The Military Conflict Considered as a Revolutionary War," in *A People Numerous and Armed: Reflections on the Military Struggle for American Independence,* rev. ed. (Ann Arbor: University of Michigan Press, 1990).
3. Richard H. Kohn, *Eagle and Sword: The Federalists and the Creation of the Military Establishment in America* (New York: Free Press, 1975), p. 45. Kohn estimates that the states had a total of 400,000 men enrolled in the militia.

4. *Militia Act of 1792: An Act more effectually to provide for the National Defence by establishing a Uniform Militia throughout the United States,* ch. 33, sec. 1, *Statutes at Large of the United States of America* 1:271–74 (1846).
5. Martha Derthick, *The National Guard in Politics* (Cambridge: Harvard University Press, 1965), p. 15., also Russell F. Weigley, *History of United States Army* (New York: Macmillan, 1967), pp. 156–57.
6. In slave-holding areas, a greater percentage of the white male population participated in the organized militia. However, it formed more a posse against the threat of slave insurrection than a force to guard against invasion.
7. U.S. Const. art. I, sec. 8.
8. *Volunteer Act of 1806: An Act authorizing a detachment from the Militia of the United States,* ch. 32, sec. 2–5, *United States Statutes at Large* 2:383–84 (1845). This was the first act authorizing the president to call for Volunteer corps from the states.
9. Frederick P. Todd, "Our National Guard," *Military Affairs* 5 (Summer, 1941): 73–86, 152–70. Although Todd's article was written before the United States entered World War II, it remains one of the clearest and most correct essays on the constitutional underpinnings of the National Guard.
10. Col. Emmons Clark, *History of the Seventh Regiment of New York, 1806–1889* (New York: n.p., 1890), 1:105.
11. "The Organized Militia of the United States," War Department Document #32 (Washington, D.C.: Government Printing Office, 1897), p. 283.
12. U.S. Volunteers were regiments similar to the state Volunteers that were mustered into federal service, but they were formed at the federal rather than state level.
13. Robert S. Chamberlain, "The Northern State Militia," *Civil War History* 4 (June, 1958): 105–9.
14. John K. Mahon, *History of the Militia and National Guard* (New York: Macmillan, 1983), pp. 108–14.
15. Derthick, *National Guard in Politics,* pp. 15–16.
16. Jerry M. Cooper, *The Army and Civil Disorder: Federal Military Intervention in Labor Disputes, 1877–1900* (Westport, Conn.: Greenwood Press, 1980), pp. 43–91.
17. Edward M. Coffman, *The Old Army: A Portrait of the American Army in Peacetime, 1784–1898* (New York: Oxford University Press, 1986), pp. 269–72; also, Robert W. Coakley, *The Role of Federal Military Forces in Domestic Disorders 1789–1878,* Army Historical Series (Washington, D.C.: Center of Military History, 1988).
18. Although Emory Upton's *The Military Policy of the United States* (Washington, D.C.: Government Printing Office, 1903) would not be published until 1904, copies of the manuscript passed through the War Department during the decades before the turn of the century. See Stephen Ambrose, *Upton and the Army* (Baton Rouge: Louisiana State University Press, 1964). The final chapter, "Influence," traces some of the effects of

Upton's unpublished work on the army from his death in March, 1881, until the Great War.

19. Derthick, *National Guard in Politics*, p. 26.
20. Jim Dan Hill, *The Minuteman in War and Peace: A History of the National Guard* (Harrisburg: Stackpole, 1964), p. 328.
21. In addition to the National Guard, a few states also maintained a naval militia as part of their organized militia.
22. *Militia Act of 1903: An Act to Promote the efficiency of the militia and for other purposes*, ch. 196, sec. 1, *Statutes at Large of the United States of America* 32:774–80 (1904).
23. Ibid., sec. 18. This part of the act stipulated that each unit would hold not less than twenty-four drills per year. Drill normally consisted of a weekly two-hour period on a weekday evening.
24. Ibid.
25. Ibid., sec. 3.
26. Section 7 of the *Militia Act of 1903* seems to dispute this. But the *Militia Act of 1908* (*An Act to further amend the Act entitled "An Act to Promote the efficiency of the militia , and for other purposes,"* ch. 204, sec. 43, *Statutes at Large of the United States of America* 35:399–403 [1909]) supports the notion of federal control only at the governor's consent.
27. *Militia Act of 1903*, sec. 5.
28. Ibid., sec. 8. The law stated that only militia officers could serve on courts-martial for officers and men of the National Guard for offenses while in federal service.
29. *Militia Act of 1908*, sec. 5. This change gave the president the power to keep militiamen in federal service as long as the president required rather than the nine-month limit of the *Militia Act of 1903*.
30. *Militia Act of 1908*, sec 5.
31. George Kearney, ed., *Official Opinions of the Attorneys General of the United States Advising the President and Heads of Departments in Relation to Their Official Duties* (Washington, D.C.: Government Printing Office, 1913), 29:322–29.
32. Derthick, *National Guard in Politics*, pp. 28–32.
33. *Volunteer Act of 1914: An Act To Provide for the raising the volunteer forces of the United States in time of actual or threatened war*, ch. 71, sec. 3, *Statutes at Large of the United States of America* 38:347 (1915).
34. John Clifford, *The Citizen Soldiers: The Plattsburg Training Camp Movement, 1913–1920* (Lexington: University Press of Kentucky, 1972), p. 118.
35. Arthur S. Link, *Wilson: Confusions and Crises, 1915–1916* (Princeton: Princeton University Press, 1964), p. 18.
36. Ibid., p. 21.
37. Ibid., pp. 30–33.
38. Ibid., p. 51.
39. Ibid., pp. 125–27.
40. Ibid.

41. *Morrill Act: An Act donating Public Lands to the several States and Territories which may provide Colleges for the Benefit of Agriculture and the Mechanic Arts,* ch. 130, sec. 4, *Statutes at Large of the United States of America,* vol. 13:504–6 (1866).
42. Gene M. Lyons and John W. Masland, "The Origins of the ROTC," *Military Affairs* 23 (Spring, 1959): 1–12.
43. John Listman, "Old Dominion's Wartime Maneuvers Along the Rio Grande," *National Guard Magazine* 46 (May, 1992): 45.
44. *National Defense Act of 1916: An Act for making further and more effectual provisions for the National Defense, and for other purposes,* ch. 134, sec. 2, *Statutes at Large of the United States of America* 39:166–217 (1917).
45. Ibid., sec 3.
46. Clifford, *Citizen Soldiers,* p. 49.
47. *National Defense Act of 1916,* sec. 57.
48. The president received the authority to decide the size of the National Guard for the territories and the District of Columbia.
49. *National Defense Act of 1916,* sec. 62.
50. Ibid., sec. 81. This section states that the National Militia Board was created by the Act of May 27, 1908, amending section 20 of the Act of January 21, 1903.
51. Ibid.
52. Ibid., secs. 25, 36. Section 25 covers the extra officers, and section 36 covers the extra sergeants.
53. Ibid., sec. 197.
54. U.S. Const. sec. 8., art. 12, and sec. 2, art. 1.
55. As an example of the ambiguity, the Chief of the Militia Bureau on April 5, 1917, sent a letter to all adjutants general (except Nevada, which had no National Guard) informing them that any National Guard officers who had not taken the federal oath would be dropped from the roles and not accepted into federal service. It added that individuals so dropped remained in the State Organized Militia, but could not handle federal property. However, law provided that the only state organized militia would be the National Guard. From a letter in file 325.36 "Federal Service-Federal oaths 1916–1917," at Mississippi State Area Command (hereafter cited as "MS-STARC"), Jackson, Mississippi.
56. *National Defense Act of 1916,* sec 37.

Chapter 2

1. U.S. Const. art. I, sec. 8., and, art. II, sec. 2.
2. The *National Defense Act of 1916,* building upon the *Militia Act of 1903,* fundamentally altered the relationship between the National Guard and the army. Under *NDA 1916,* National Guard officers held federal commissions in addition to their state commissions, and enlisted men took a dual oath to both

their state and the federal constitutions. These changes were necessary to overcome the constitutional barriers to using militia beyond U.S. borders.

3. "Circular Letter" of June 2, 1917, from F. D. Evans of Headquarters Southeastern Department, to All Commanding Officers. Located at MS-STARC, Jackson, Miss.
4. Ibid.
5. Almost all *Annual Reports* of the period from southern, and some northern, adjutants general show several instances every year of the use of the National Guard in this manner.
6. Merle T. Cole, "The Department of Special Deputy Police, 1917–1919," *West Virginia History* 44 (Summer, 1983): 324–25, contains a passage from the legislative debates in that state over the creation of state police forces, in which the popular images of the Civil War home guard were still strong.
7. Robert Hawk, *Florida's Army: Militia/State Troops/National Guard, 1565–1985* (Englewood, Fla.: Pineapple Press, 1986), p. 138.
8. Ibid., pp. 140–41.
9. Ibid.
10. George Cassel Bittle, "In Defense of Florida: The Organized Florida Militia from 1821 to 1920" (Ph.D. diss., Florida State University, 1965), p. 426.
11. Ibid.
12. Ibid.
13. Ibid., p. 427.
14. Hawk, *Florida's Army*, p. 143.
15. Bittle, "In Defense of Florida," p. 427.
16. Counties acknowledging the formation of these forces under their sheriffs were Sumter, Carrollton, Pickens, Hamilton, Marion, Limestone, Randolph, Colbert, Russell, Monroe, Shelby, and Morgan. These twelve were 8 percent of the sixty-seven counties in the state.
17. Governor Henderson to Alabama county sheriffs, June 5, 1917, folder "National Guard," Records Group RCA: G122, Alabama State Archives (hereafter cited as "ASA"), Montgomery.
18. B. L. Malone to Governor Henderson, Sept. 9, 1917, folder "National Guard," Records Group RC2: G122, ASA, Montgomery.
19. J. C. Bowie to Governor Henderson, including clipping from unidentified newspaper, June 12, 1917, RC2: G122, ASA, Montgomery.
20. Edward M. Coffman, *The War to End All Wars: The American Military Experience in World War I* (New York: Oxford University Press, 1968), pp. 26–27.
21. A. H. Powell, owner of "The Variety Store" in Hackleburg, Alabama, on behalf of A. G. Freemen, to Governor Henderson, June 27, 1917, papers of Governor Henderson, ASA, Montgomery.
22. Uniform Rank, Woodsmen of the World, to Governor Henderson, July 19, 1917, papers of Governor Henderson, ASA, Montgomery.
23. Governor Henderson to the Uniform Rank, Woodsmen of the World, July 27, 1917, papers of Governor Henderson, ASA, Montgomery.

24. J. W. Cantrell, a partner at J. J. Lawler & Co., Dealers in General Merchandise, of Hodges, Alabama, to Governor Henderson, Sept. 21, 1917, papers of Governor Henderson, ASA, Montgomery.
25. This was part of a larger trend from the period of increased tension both before and after the declaration of war in which many communities from around the state requested permission to raise National Guard units for service in the war. After the declaration, the adjutant general replied to the inquiries by saying that uncertainty about changes in War Department policies prevented him from authorizing new units. After April 12, 1917, he replied by stating War Department policies for the creation of new National Guard units. From file "325.4 Organizations-New Units Re-organization of Units, Assignments, 1917," at MS-STARC, Jackson.
26. W. W. Robertson, Postmaster of McComb, Mississippi, to Brig. Gen. Erie C. Scales, Apr. 6, 1917, folder "325.4 Home Guards Boy Scouts," at MS-STARC, Jackson.
27. Brig. Gen. Erie C. Scales, Adjutant General of Mississippi to Mr. W. W. Robertson of McComb, Mississippi, Apr. 7, 1917, folder "325.4 Home Guards Boy Scouts," at MS-STARC, Jackson.
28. R. T. Luke to Brig. Gen. Erie C. Scales, Apr. 7, 1917, folder "325.4 Home Guards Boy Scouts," at MS-STARC, Jackson.
29. C. E. Anding to Brig. Gen. Erie C. Scales, Apr. 9, 1917, folder "325.4 Home Guards Boy Scouts," at MS-STARC, Jackson.
30. W. A. Hickman to Brig. Gen. Erie C. Scales, Apr. 9, 1917, folder "325.4 Home Guards Boy Scouts," at MS-STARC, Jackson.
31. Mr. Wallace B. Cameron of Meridian, Mississippi, to Brig. Gen. Erie C. Scales, Apr. 11–17, 1917, and Scales's reply of Apr. 17, 1917, folder "325.4 Home Guards Boy Scouts," at MS-STARC, Jackson.
32. Knights of Pythias of Mississippi to Brig. Gen. Erie C. Scales, Apr. 9, 1917, folder "325.4 Home Guards Boy Scouts," at MS-STARC, Jackson.
33. Letter of unknown authorship to Brig. Gen. Erie C. Scales, Apr. 9, 1917, folder "325.4 Home Guards Boy Scouts," at MS-STARC, Jackson.
34. W. W. Gwin, Mayor of Tchula, to Brig. Gen. Erie C. Scales, Apr. 26, 1917, folder "325.4 Home Guards Boy Scouts," at MS-STARC, Jackson.
35. S. R. Young to Brig. Gen. Erie C. Scales, Apr. 19, 1917, folder "325.4 Home Guards Boy Scouts," at MS-STARC, Jackson.
36. Brig. Gen. Erie C. Scales, *Biennial Report of the Adjutant General State of Mississippi from January 1, 1916–December 31, 1917* (Jackson, Miss.: Tucker Printing, 1917), p. 132.
37. Ibid., pp. 3–5.
38. *Daily Clarion-Ledger* (Jackson) carried several stories on this theme: "Biloxi Fears Menace of Teutonic Spies / Alderman claims spies have visited, guards needed for waterworks, electric, gas plants, bridge," Apr. 5, 1917, p. 4. Also, Germans and Mexicans "Arrange to Attack along the Border," quotes reports of Germans drilling Villa's troops, Apr. 6, 1917, p. 5.

39. "Meridian fears Negroes," headline for story on reports of German agitators among the blacks. A near riot erupted when a white man told a black man that he ought to go to Mexico if he wanted to fight for the Germans. The paper also contained stories that discounted the possibility of Germans fomenting black unrest and mentioned that social inequality probably played a large role in unrest among the blacks of the South. *Daily Clarion-Ledger,* Apr. 10, 1917, p. 4.
40. *Daily Clarion-Ledger,* Apr. 11, 1917, p. 6.
41. Vicksburg, on the other side of the state, was the scene of the worst racial unrest during the war. Neil R. McMillen, *Dark Journey: Black Mississippians in the Age of Jim Crow* (Chicago: University of Illinois Press, 1989), pp. 304–305.
42. In a telegram to the Commanding General, USA, at Fort Sam Houston, General Scales said that the First Mississippi Infantry was in federal service stationed at Jackson, Mississippi, with 75 percent of it on sentry duty within the state. From folder "325.36 Federal Service-Federal Oaths 1916–1917," at MS-STARC, Jackson.
43. Maj. Gen. Leonard Wood at Headquarters Southeastern department in Charleston, South Carolina, to the "Honorable Thoe [*sic*] G. Bilbo," June 13, 1917, folder "325.36 Federal Service-Federal Oaths 1916–1917," at MS-STARC, Jackson.
44. Ibid.
45. Ibid.
46. Sixty-Fifth Congress, 1st sess. Ch. 28. June 14, 1917. In the Army Appropriation Bill for the year ending June 30, 1919, Congress authorized the payment of $2,500,000 to procure the arms and equipment authorized under the Act of June 14, 1917. 65th Congress, 2nd sess. Ch. 143.
47. Brig. Gen. Erie C. Scales, *Biennial Report of the Adjutant General of Mississippi for the Years 1918–1919 to the Governor* (Jackson: Tucker Printing House, 1920), p. 132.
48. Ibid.
49. Brig. Gen. Eric C. Scales to the Acting Chief of Ordnance, United States Army, Mar. 11, 1918, folder "325.4 Home Guards Boy Scouts," at MS-STARC, Jackson.
50. Scales, *Adjutant General Report for 1916–1917*, pp. 132–33.
51. "Circular Letter Number 17," from the Chief of the Militia Bureau (hereafter cited as CMB), Aug. 24, 1917, in the *Annual Report of the Chief of the Militia Bureau 1918* (hereafter cited as *ARCMB*) (Washington, D.C.: Government Printing Office, 1919), p. 1126.
52. *ARCMB 1918*, p. 1127.
53. Ibid., p. 1128.
54. "Circular Letter #3," War Department, from "J. McO. Carter," major general, National Army. Quoted in Harvey J. Moss, *Report of the Adjutant General of Washington, 1917–1918* (Olympia: Frank M. Lamborn, Public Printer, 1918), p. 4.
55. *ARCMB 1918*, pp. 1128–29.

56. *ARCMB 1918*, p. 1129. These are the states listed in the report, although it does not include all states that took this option.
57. Ibid., p. 1108.
58. Jim Dan Hill, *The Minute Man in War and Peace: A History of the National Guard* (Harrisburg, Pa.: Stackpole, 1964), p. 328.
59. Martha Derthick, *The National Guard in Politics* (Cambridge, Mass.: Harvard University Press, 1965), pp. 22–24.
60. Hill, *Minuteman in War and Peace*, p. 328.
61. *ARCMB 1918*, p. 1108.
62. Valentine J. Belfiglio, *Honor, Pride, Duty: A History of the Texas State Guard* (Austin: Eakin Press, 1995), p. 54.
63. John Glendower Westover, "The Evolution of the Missouri Militia, 1804–1919." (Ph.D. diss., University of Missouri, 1948), p. 254.
64. Harvey C. Clark, Adjutant General, *Report of the Adjutant General of Missouri January 1, 1917–December 31, 1920* (Jefferson City: State of Missouri, 1920), p. 37.
65. Christopher C. Gibbs, *The Great Silent Majority: Missouri's Resistance to World War I* (Columbia: University of Missouri Press, 1988), p. 61.
66. Ibid.
67. Westover, "Evolution of the Missouri Militia," pp. 254–55.
68. Clark, *Report of January 1, 1917–December 31, 1920*, p. 35.
69. Gibbs, *Great Silent Majority*, p. 62.
70. Clark, *Report of January 1, 1917–December 31, 1920*, p. 37.
71. Ibid., p. 38.
72. Westover, "Evolution of the Missouri Militia," p. 254.
73. Gibbs, *Great Silent Majority*, p. 62.
74. Clark, *Report of January 1, 1917–December 31, 1920*, p. 38.
75. Gibbs, *Great Silent Majority*, p. 62.
76. Clark, *Report of January 1, 1917–December 31, 1920*, p. 39.
77. Lawrence O. Christensen, "World War I in Missouri," *Missouri Historical Review* 90 (July, 1996): 427.
78. Clark, *Report of January 1, 1917–December 31, 1920*, p. 23.
79. Westover, "Evolution of the Missouri Militia," p. 255.
80. Ibid., p. 254.
81. Clark, *Report of January 1, 1917–December 31, 1920*, pp. 24–25.
82. Ibid., pp. 25–26.
83. Ibid., p. 35.
84. Gibbs, *Great Silent Majority*, p. 63.
85. Ibid., p. 60.
86. Christensen, "World War I in Missouri," p. 428.
87. Gibbs, *Great Silent Majority*, p. 60.
88. Eben Putnam, ed., *Report of the Commission on Massachusetts's Part in the World War History*, vol. 1, (Boston: Commonwealth of Massachusetts, 1931), p. 160.
89. Message of Mar. 22, 1917, from Governor McCall to the General Court. Included in *Public Document No. 49, Fourteenth Annual Report of the Police*

Commission for the City of Boston: Year Ending November 30, 1919 (Boston: Wright and Potter Printing Co., State Printers, 1920).

90. Ibid.
91. Chapter 148 of the Acts of 1917, "To Provide for the Organization of a Home Guard in Time of War," approved Apr. 5, 1917.
92. Ibid., Section 1.
93. Ibid., Section 6.
94. *History of the Richardson Light Guard of Wakefield, Mass.: Covering the Third Quarter-Century Period 1901–1926* (Wakefield, Mass.: Item Press, 1926), pp. 81–82.
95. Ibid., p. 96.
96. "A Sketch of the Concord Company, Massachusetts State Guard. 1917–1921," p. 2. From "Collection of Materials Relating to the Concord Company of the Massachusetts State Guard," at the Concord (Mass.) Free Public Library.
97. "Scrap Book Events in Concord, Vol. II 1915–1921," at the Concord (Mass.) Free Public Library, p. 16.
98. Ibid.
99. "Sketch of the Concord Company," p. 2.
100. Ibid., p. 3.
101. Charles S. Zack, *Holyoke in the Great War* (Holyoke, Mass.: Transcript Publishing Company, 1919), p. 293.
102. Ibid., p. 295.
103. Ibid.
104. Putnam, *Report of the Commission on Massachusetts's Part in the World War,* 1:53.
105. Ibid.
106. Ibid., p. 54.
107. Eben, *Massachusetts's Part in the World War,* 1:162.
108. United States Guards were a part of the army formed from men who had been conscripted and trained but who had later been declared unfit for overseas service. For a further discussion of the U.S. Guards, see, Merle T. Cole, "United States Guards, United States Army," in *Military Collector and Historian: Journal of the Company of Military Historians* 40 (Spring, 1988): 2–5.
109. Chapter 188 of the Acts of 1918, approved May 2, 1918.
110. *The Two Hundred and Seventy-Ninth Annual Record of the Ancient and Honorable Artillery Company of Massachusetts, 1916–1917* (Boston: George E. Crosby Co., 1918), p. 65.
111. Ibid., p. 43.
112. Ibid.
113. Ibid., p. 44.
114. State of Rhode Island and Providence Plantations, *Annual Report of the Adjutant General and Quartermaster General of the State of Rhode Island for the Year 1917* (Providence R.I.: Oxford, [1918]), p. 9.

115. State of Rhode Island and Providence Plantations, *Annual Report of the Adjutant General and Quartermaster General of the State of Rhode Island for the Year 1918* (Providence, R.I.: E. L. Freeman Company, [1919]), p. 12.
116. Rhode Island, *Annual Report for the Year 1917*, p. 9.
117. Ibid., p. 10.
118. Ibid., pp. 33–37.
119. Rhode Island, *Annual Report for the Year 1918*, p. 3.
120. Ibid., p. 4.
121. Ibid., The Act refers to the force as the "State guard," but the adjutant general consistently uses "State Guard."
122. Ibid., p. 6.
123. Ibid.
124. Ibid., p. 5.
125. The 103 represented three officers, with the other hundred being a supply sergeant, NCOs, cooks, and men.
126. Rhode Island, *Annual Report for the Year 1918*, p. 6.
127. Ibid., p. 7.
128. Ibid., p. 8.
129. Ibid., p. 9.
130. Rhode Island, *Annual Report for the Year 1918*, p. 10.
131. Ibid., pp. 17–19.
132. This is the number given in the U.S. National Guard Bureau, *Annual Report of the Chief of the National Guard Bureau 1941* (Washington, D.C.: Government Printing Office, 1946), p. 32. However, an exact number cannot be given due to the nonstandard methods in each state for deciding what constituted a home guard.
133. Ibid.
134. Ibid.

Chapter 3

1. National Guard Association, *Proceedings of the Convention of the National Guard Association Held at Richmond, Virginia November 14 and 15 1918* (n.p., n.d.).
2. Ibid.
3. Ibid.
4. Ibid.
5. The National Guard Reserve consisted of men who were in the National Guard but did not attend drill or Annual Training. They did, however, maintain a commitment to respond in the event of a state or federal mobilization. This name would later be used by Washington State and several other states for its home guard after World War II.
6. Moss, *Report of 1917–1918*, p. 19.
7. Ibid.

8. Ibid., p. 20.
9. State of Washington Military Department, Office of the Adjutant General, *A Brief History of the National Guard of Washington* (n.p.: State of Washington, 1957), p. 16.
10. Moss, *Report of 1917–1918*, p. 24.
11. State of Washington, *Brief History of the National Guard of Washington*, p. 16.
12. Moss, *Report of 1917–1918*, p. 25.
13. Ibid., p. 6.
14. State of Washington, *Brief History of the National Guard of Washington*, p. 32.
15. Moss, *Report of 1917–1918*, p. 4.
16. State of Washington, *Brief History of the National Guard of Washington*, p. 7.
17. Moss, *Report of 1917–1918*, p. 5.
18. National Guard Association, *Proceedings of the Convention of the National Guard Association Held at Richmond, Virginia November 14 and 15 1918* (n.d., n.p.), pp. 1–5.
19. State of Washington, *Brief History of the National Guard of Washington*, p. 32.
20. Ibid., p. 32.
21. *Sixteenth Biennial Report of the Adjutant General of the State of Oregon To the Governor and Commander-in-Chief For the Period November 1, 1916, to October 31, 1918* (Salem: State Printing Department, 1919), p. 18.
22. Ibid., p. 19.
23. Ibid.
24. Ibid.
25. The term "State Guard" is used here in its generic sense of an organized militia with an obligation for use anywhere within the state but having no federal mission. The adjutant general of Oregon referred to the force as a State Guard, although its proper name was the "Oregon Guard." In this usage, Oregon followed the practice of New York, which referred to its wartime militia as the "New York Guard," and more informally as the "State Guard."
26. *Sixteenth Biennial Report of the Adjutant General of the State of Oregon*, p. 20.
27. Ibid., p. 21.
28. Ibid., p. 22.
29. Ibid., p. 21.
30. Ibid.
31. Ibid., p. 24.
32. Ibid., p. 23.
33. Ibid., p. 24.
34. *Seventeenth Biennial Report of the Adjutant General of the State of Oregon To the Governor and Commander-in-Chief For the Period November 1, 1918, to October 31, 1920* (Salem: State Printing Department, 1921), p. 9.
35. Ibid., p. 10.
36. Governor Henderson to Virgil V. Evans, Nov. 8, 1918, folder "National Guard," Records Group RC2: G122, ASA, Montgomery.

37. Virgil V. Evans to Governor Henderson, Nov. 9, 1918, Records Group RC2: G122, folder "National Guard," ASA, Montgomery.
38. This announcement simply reiterated "Circular Letter Number 17" of August 24, 1917, from the CMB, which had outlined the policy of the Secretary of War on new National Guard units organized after the draft of the National Guard on August 5, 1917.
39. Governor Henderson to the Acting Chief of the Militia Bureau, Oct. 15, 1918, Records Group RC2: G122, folder "National Guard," ASA, Montgomery.
40. *ARCMB 1920*, pp. 5–7.
41. "325.44 Strength-Returns 1918-19-20-21-22," at MS-STARC, Jackson.
42. U.S., Congress, Senate, *A Bill to authorize the issue to States and Territories and the District of Columbia of rifles, pistols, machine guns, and other property for the equipment of home guards.* S.3289, 66th Congress, 1st sess., 1919. The bill was introduced on October 22, (calendar day, October 23), in the Senate, read twice, and referred to the Committee on Military Affairs.
43. Brig. Gen. Erie C. Scales to Sen. Pat Harrison, Oct. 27, 1919, folder 325.4 "Home Guards Boy Scouts," at MS-STARC, Jackson.
44. Ibid.
45. Ibid.
46. In *Dark Journey*, p. 304, McMillen explains that whites in Mississippi were fearful of black veterans returning expecting change or "corrupted" by service with the French. As result, the years 1917–19 were "among the most violent and restrictive since the last years of Reconstruction."
47. Ibid.
48. Sen. Pat Harrison of Mississippi to the Adjutant General of Mississippi, Brig. Gen. Erie C. Scales, Nov. 3, 1919, folder "325.4, Home Guards Boy Scouts" at MS-STARC, Jackson.
49. Brig. Gen. Erie C. Scales to Sen. Pat Harrison, Nov. 13, 1919, folder "325.4 Home Guards Boy Scouts," at MS-STARC, Jackson.
50. Ibid.
51. The New York Guard contained, at its largest, four brigades of Infantry, with a total of fourteen regiments. It also contained one regiment of cavalry, one separate squadron of cavalry, two regiments of field artillery, one regiment of engineers, one machine gun battalion, plus supporting elements. Brigadier General Charles W. Berry, *Annual Report of the Adjutant General for the Year 1919* (Albany: J. B. Lyons Company, 1921), pp. 3, 81.
52. Ibid., p. 11.
53. Ibid., p. 3.
54. Ibid., p. 4.
55. Ibid., p. 20.
56. *Biennial Report of the Adjutant General of the State of Delaware for the Two Years Ending December 31, 1918* (Milford, Dela.: Milford Chronicle Publishing Co., [1919]).
57. Ibid.

58. *Biennial Report of the Adjutant General of the State of Delaware for the Two Years Ending December 31, 1920* (Milford, Del.: Milford Chronicle Publishing Co, [1921]).
59. Merle T. Cole. "Second Regiment, Infantry, Maryland State Guard: An Early State Defense Force," *Journal of Military History* 53 (Jan., 1989): 23.
60. The Maryland National Guard at that time consisted of one brigade comprised of the First, Fourth, and Fifth Maryland Infantry Regiments.
61. Cole, "Second Regiment, Infantry, Maryland State Guard," p. 24.
62. *Report of the Adjutant General State of Maryland for 1916–1917* (Baltimore: King Brothers State Printers, [1918]), p. 15.
63. Cole, "Second Regiment, Infantry, Maryland State Guard," p. 28.
64. Lieut. C. Arthur Eby, "A Record of the Maryland State Guard," (Baltimore: 1920), pp. 1–2 (not paginated), at the Army War College, Carlisle Barracks, Pa.
65. Cole, "Second Regiment, Infantry, Maryland State Guard," p. 28.
66. Ibid., p. 16.
67. Merle T. Cole, "Maryland's State Defense Force," *Military Collector and Historian* 39 (Winter, 1987): 152.
68. Cole, "Second Regiment, Infantry," p. 31.
69. Eby, "Record of the Maryland State Guard," p. 2.
70. "General Order Number 16," from the War Department, quoted in Maryland, Council of Defense, *Report of the Maryland Council of Defense to the Governor and the General Assembly, 1919–1920* (n.p., 1920), p. 19.
71. Cole, "Second Regiment, Infantry, Maryland State Guard," p. 29.
72. Eby, "Record of the Maryland State Guard," pp. 3–8.
73. Ibid.
74. Ibid.
75. Cole, "Second Regiment, Infantry, Maryland State Guard," p. 31.
76. Ibid., p. 30.
77. Ibid.
78. Ibid., p. 31.
79. Ibid.
80. Eby, "Record of the Maryland State Guard."
81. Cole, "Second Regiment, Infantry, Maryland State Guard," p. 31.
82. *Report of the Adjutant General State of Maryland for 1918–1919* (Annapolis: Weekly Advertizer, n.d.), p. 8.
83. *Biennial Report of the Adjutant General State of Tennessee, Nashville, March 19, 1917* (n.p., [1917]), p. 10.
84. Ibid., p. 11.
85. Ibid., p. 12.
86. *Biennial Report of the Adjutant General of the State of Tennessee for the Period January 1, 1917 to January 1, 1919* (n.p., [1926]), p. 26.
87. Ibid., p. 28.
88. Ibid., p. 20.

89. Ibid., p. 36.
90. Ibid., p. 40.
91. James Baca, Adjutant General, *Report of the Adjutant General of the State of New Mexico from November 30, 1916 to November 30, 1918* (n.p., n.d.). This report makes no mention of any home guard organizations.
92. Ibid.
93. Charles Springer, Chairman Executive Committee, New Mexico Council of Defense, *Report of the Council of Defense of the State of New Mexico Nineteen Hundred Eighteen* (n.p.: New Mexico Council of Defense, n.d.), p. 36.
94. Ibid.
95. Ibid.
96. Ibid., p. 37.
97. National Guard Association, *Proceedings of the Convention of the National Guard Association Of The United States Saint Louis Missouri May fifth and sixth Nineteen-nineteen* (n.p., n.d.), pp. 14–17.
98. Ibid., pp. 17–20.
99. *State of Connecticut Report of the Adjutant General Two Years ended September 30, 1918* (Hartford: Published by the State, 1919), p. 13.
100. Lloyd W. Fowles, *An Honor to the State: The Bicentennial History of the First Company Governor's Foot Guard* (Hartford: First Company Governor's Foot Guard, 1917), pp. 138–39; also, William J. Prendergast, compiler, *Two Hundred Years: The Second Company Governor's Foot Guard, 1775–1975* (n.p.: Second Company Governor's Foot Guard, 1975), p. 29.
101. "Section 61" of the *National Defense Act of 1916* forbade the states from keeping an organized militia not part of the National Guard or naval militia. Although these organizations were technically part of the organized militia of Connecticut, they were never used in calculations of representatives the state sent to the National Guard Association.
102. "State Guard to be Recruited to its Full Quota," *Veterans' Journal and State Guard News*, Apr., 1920, p. 15.
103. Ibid.
104. Francis Russell, *A City in Terror: 1919—The Boston Police Strike* (New York: Viking Press, 1975,) pp. 48, 99.
105. "Public Document No. 49," *Fourteenth Annual Report of the Police Commission for the City of Boston: Year Ending November 30, 1919* (Boston: Wright & Potter Printing Co., State Printers, 1920), p. 18.
106. Ibid., p. 5.
107. Ibid., p. 18.
108. Ibid., p. 19.
109. Eben, *Massachusetts's Part in the World War*, p. 163.
110. Commissioner Curtis to Mayor Peters, Sept. 10, 1919. In the *Report of the Citizen's Committee Appointed By Mayor Peters to Consider the Police Situation.* Included in the *Fourteenth Annual Report of the Police Commissioner for the*

City of Boston: Year Ending November 30, 1919 (Boston: Wright & Potter Printing Co., State Printers, 1920).

111. Ibid., Governor Coolidge's Proclamation, Sept. 11, 1919.
112. "Sketch of the Concord Company," p. 8.
113. *History of the Richardson Light Guard of Wakefield, Mass.: Covering the Third Quarter-Century Period 1901–1926* (Wakefield, Mass.: Item Press, 1926), p. 105.
114. "Sketch of the Concord Company," p. 9.
115. Eben, *Massachusetts's Part in the World War,* p. 164.
116. Francis, *City in Terror,* p. 5.
117. Ibid.
118. "Sketch of the Concord Company," p. 15.
119. Ibid., p. 16.
120. Ibid.

Chapter 4

1. U.S. National Guard Bureau, *Annual Report of the Chief of the National Guard Bureau 1934* (hereafter cited as *ARCNGB*) (Washington, D.C.: Government Printing Office, 1934), p. 11.
2. *ARCNGB 1940*, p. 14.
3. Ibid., p. 15.
4. Ibid., p. 19.
5. Executive Order No. 8530.
6. Public Law 213, 77th Congress, approved Aug. 18, 1941, known as the "Service Extension Act." This law applied to men conscripted under Selective Service as well as National Guardsmen.
7. *ARCNGB,* pp. 16–17.
8. Mahon, *History of the Militia and National Guard,* p. 179.
9. *ARCNGB,* p. 32. This figure is approximate because of questions about what constitutes an organized militia. For example, California simply licensed men to bear arms and referred to these men collectively as a home guard, yet the state is not counted in the War Department reports as having created militia forces in the Great War.
10. Ibid.
11. Ibid. Also, see Merle T. Cole, "United States Guards, United States Army," *Military Collector and Historian* 40 (Spring, 1988): 2–5.
12. The term "Minute Men" for these proposed forces should not be confused with the use of the same term for a force many states, such as Pennsylvania, created after Pearl Harbor. These forces, which the National Rifle Association advocated, were adjuncts to the State Guards but were usually self-armed and liable for local defense only. Other states had similar forces but usually called them State Guard Reserve.
13. Ben M. Van Hook of Missouri, of the American Legion, to Col. Paul V.

McNutt of the SEC Administration, Sept. 15, 1939, folder "Plan Minute Men of Am," RG 165, box 470, National Archives I, Washington, D.C.

14. Bill of Feb. 12, 1941, in the House for a "Home Defense Organized Reserve," folder 032.1, box 104, RG 168, National Archives II, College Park, Md.
15. Ibid.
16. Memorandum for Chief of Staff on Internal Security of June 17, 1940, folder "Civil Defense Branch," box 473, RG 165, National Archives I, Washington, D.C. The October, 1875, Supreme Court case *U.S. v. Cruikshank,* (92 US 542) concerned a conspiracy to violate the voting rights of two black men in Louisiana. The Supreme Court, in its ruling, wrote that although the United States government was supreme vis-à-vis the state governments, it could "not grant nor secure rights to citizens" not expressed nor implied under its protection, and that "[s]overeignty, for the protection of the rights of life and personal liberty within the respective States, rests alone with the States."
17. "No State shall, without the consent of Congress, . . . keep troops, or ships of war in time of peace . . ." U.S. Const. art. I, sec 10.
18. June 10, 1940, proposal by Raymond J. Kelly, National Commander, the American Legion, folder "Opinions of JAG," box 470, RG 165, National Archives I, Washington, D.C.
19. Memorandum, Aug. 27, 1940, George C. Marshall on the "American Legion Service of Security Plan," folder 42494, box 472, RG 165, National Archives I, Washington, D.C.
20. Alabama State Military Department, *Quadrennial Report of the Adjutant General* (1942), pp. 75–78.
21. The Military Code of Louisiana, to cite just one example, includes the provision that "[n]o civil Organization, Society, Club, Post, Order Fraternity, Association, Brotherhood, Body Union, League, or other combination of persons or Civil Group shall be enlisted in the Louisiana State Guard as an organization or unit." Most other states had similar stipulations. The wording came from "section 8" of the federal government's *Model State Guard Act,* which states were encouraged to adopt.
22. Memorandum, Sept. 5, 1940, Col. J. M. Churchill of the General Staff, folder 42494, box 472, RG 165, National Archives I, Washington, D.C.
23. Stetson Conn, *Guarding the United States and Its Outposts* (Washington, D.C.: Office of the Chief of Military History, Dept. of the Army, 1964), p. 75.
24. Memorandum, Col. J. M. Churchill of the General Staff, Sept. 5, 1940, folder 42494, box 472, RG 165 National Archives I, Washington, D.C.
25. Memorandum, July 19, 1940, from J. M. Churchill, General Staff, Chief, Civil Defense Branch, to Colonel Camp, folder 42494, box 472, RG 165 National Archives I, Washington, D.C.
26. Memorandum for Assistant Chief of Staff, Aug. 2, 1940, folder 42494, box 472, RG 165, National Archives I, Washington, D.C.
27. Memorandum for the Chief of Staff, Sept. 18, 1940, box 473, file 43687, RG 165, National Archives I, Washington, D.C.

28. See Richard G. Stone, Jr., *A Brittle Sword: The Kentucky Militia, 1776–1912* (Lexington: University Press of Kentucky, 1977), for an example of this practice. Throughout the late nineteenth century, the organized militia of Kentucky, which would evolve into the National Guard, was referred to as "the State Guard," and the practice would continue informally in the twentieth century.
29. For more information on the British Home Guard, see S. P. MacKenzie, *The Home Guard: A Military and Political History* (New York: Oxford University Press, 1995).
30. Colonel Elbridge Colby, "Report on the British Home Guards," folder 324.5 "Home Guard File," box 470, RG 165, National Archives I, Washington, D.C. This same report appeared in the *Annual Report of the Chief National Guard Bureau* for 1946, Appendix C.
31. The bill had originally been introduced into Congress on July 1, 1940.
32. *ARCNGB 1941*, p. 34. State forces originally received the .30 caliber M-1917 Enfield with accessories.
33. Ibid., p. 35.
34. Ibid, p. 36.
35. Army Regulation No. 850-250, issued Apr. 21, 1941. At the Army War College Library, Carlisle Barracks, Pa.
36. "C.Q.S." to Colonel Grimes, July 17, 1940, folder 42494, box 472, RG 165, National Archives I, Washington, D.C.
37. See Emory Upton, *The Military Policy of the United States* (Washington, D.C.: Government Printing Office, 1907). Although published posthumously in 1903, the manuscript had been circulating around the War Department for many years. In it, Upton claimed that one of the chief weaknesses of the American military was the interference of the states. He proposed as a remedy a completely federal reserve for the army.
38. *ARCNGB 1941*, p. 33.
39. T. A. Frazier, *From Peace to War: State of Tennessee Consolidated Report of the Adjutant General for the Period January 16, 1939–January 4, 1943* (n.p., n.d.), p. 21.
40. *ARCNGB 1941*, p. 17.
41. Conn, *Guarding the United States and Its Outposts*, p. 76.
42. "Status of State Guards," *Army and Navy Register* 62 (Aug. 9, 1941): 14.
43. "Status of State Guards," *Army and Navy Register* 62 (Nov. 1, 1941): 5.
44. File No. SI-1123, Headquarters, First Corps Area, in Boston, Apr. 1, 1941. folder 42494, box 472, RG 165, National Archives I, Washington, D.C.
45. *ARCNGB 1941*, p. 34.
46. Note, Apr. 15, 1941, with attached newspaper comic page, folder 42494, box 472, RG 165, National Archives I, Washington, D.C.
47. Exchange of June–July 1941, folder 324 2 . . . 9, box 116, RG 168, National Archives II, College Park, Md.
48. Folder 42494, box 472, RG 165, National Archives I, Washington, D.C.

49. S. 17315, Aug., 1941. The law as passed was Public Law 214-77th Congress. The War Department announced its passage in *Bulletin No. 29*, on Aug. 18, 1941. File 032.1., box 104, RG 168, National Archives II, College Park, Md. The *ARCNGB 1942* incorrectly reports that the initial legislation included the territories of Alaska and Hawaii and that the amendment of 18 August 1941 applied only to Puerto Rico and the Canal Zone. In reality, the new law applied to all four possessions, p. 65.
50. C. A. Davo, the Adjutant General [*sic*] of the District of Columbia, to the CNGB, Apr. 11, 1941. Folder 325 "Dist. Col. State Guard," box 134, RG 168, National Archives II, College Park, Md.
51. Estimate of CNGB Maj. Gen. John F. Williams, in 1947 folder 370.1., box 121, RG 168, National Archives II, College Park, Md.

Chapter 5

1. *ARCNGB 1941*, p. 34. The term "State Guard" is here used in the sense that the War Department used the term—for organized militia forces raised by the states to replace the National Guard in state service. Most states eventually adopted the name for their force, but a few did not. For example, some State Guard-type forces were the Kentucky Active Militia, Michigan State Troops, Missouri Military Reserve Force, New York Guard, and Illinois Reserve Militia.
2. Ibid., 1941, p. 40. These figures are suspect, however, because among the states listed is Oklahoma, with 520 authorized and 411 mustered in its "State Defence Force," yet the force never became viable and the state never created a State Guard in World War II.
3. See chapter 4, note 21.
4. Todds, "Our National Guard."
5. The right of states to prohibit private groups from assembling as militia was upheld in the Supreme Court ruling of *Presser v. Illinois* in 1886.
6. Tom Pankey of the Pankey Brokerage Company, to Governor Dixon, May 14, 1940, folder #2, Records Group SG 12242, Adjutant General, ASA, Montgomery.
7. The organization "The Christian American" of Houston, Texas, sent Governor Dixon on April 3, 1941, a copy of the *Texas Defense Guard Act* and requested a copy of similar Alabama laws. Governor Dixon's legal advisor replied on April 14, 1941, that Alabama had made no direct statutory provisions for a home guard. From folder Home Guard, Records Group SG 12261, ASA, Montgomery.
8. C. H. Day, major, 4th Field Artillery, Acting Assistant Adjutant General, Aug. 6, 1940, folder #2, Records Group SG 12242, Adjutant General, ASA, Montgomery.
9. Lieut. Gen. S. D. Embick, Headquarters 4th Corps Area, to Governor Dixon, Aug., 14, 1940, folder #2, Records Group SG 12242, Adjutant General, ASA, Montgomery.

10. Ibid.
11. Memorandum, Headquarters, 4th Corps Area, to Governor Dixon, Sept. 3, 1940. Included was the August 21, 1940, proposal from Raymond J. Kelly, the National Commandant of the American Legion, and the reply of the 4th Corps HQ to the American Legion, folder #2, Records Group SG 12242, Adjutant General, ASA, Montgomery.
12. Ibid. The August 28, 1940, reply of the 4th Corps Area was written by Col. John P. Smith, General Staff Corps Chief of Staff, on behalf of Lt. Gen. S. D. Embick.
13. Ibid.
14. During early desperate months of the Pacific war, Maj. Gen. Patch formed and commanded the Americal [*sic*] Division on New Caledonia out of National Guard units orphaned by reorganization. After the war, Lt. Gen. Patch would head the board bearing his name that laid the foundation of the army's postwar structure.
15. Alabama State Military Department Office of the Adjutant General, *Quadrennial Report of the Adjutant General for Four Year Period Ending September 30, 1942* (Montgomery: State of Alabama, 1942), p. 75.
16. Alabama State Military Department, *State Regulation Number 4*, Oct. 1, 1940, p. 1. At the Hoole Rare Book Collection, University of Alabama, Tuscaloosa.
17. Ibid., p. 3.
18. Ibid.
19. Years after the war, black posts were admitted into the American Legion in Alabama following heated debate. Much of the justification for admitting such chapters was based on the idea that all veterans benefited from the Legion's lobbying efforts and black veterans should bear some of the cost through their dues.
20. Letter of Resignation, Mr. Carey Robinson of Alexander City, Oct. 3, 1940, from SG 12253—American Legion folder, ASA, Montgomery.
21. Alabama Headquarters of the American Legion, "Bulletin to Posts," dated Oct. 14, 1940, from SG 12253—American Legion folder, ASA, Montgomery.
22. Ralph H. Collins, *Bulletin* of Jan. 15, 1941, from SG 12253—American Legion folder, ASA, Montgomery.
23. Alabama Department Headquarters of the American Legion *Memo* of June 18, 1941, to all department and post commanders, from SG 12253—American Legion folder, ASA, Montgomery.
24. State of Alabama, *Quadrennial Report*, pp. 10, 13, 139–42.
25. "Headquarters Alabama State Guard" in Montgomery, sent to post commanders and Adjutants of the American Legion in Alabama from "George L. Cleere, Colonel, Alabama State Guard," Nov. 28, 1940, from SG 12253—American Legion folder, ASA, Montgomery.
26. Ibid.
27. Representative Joseph N. Langan to Governor Dixon, Sept. 11, 1940, folder Home Guard, Records Group SG 12261, ASA, Montgomery.

28. Reply, Governor Dixon to Rep. Langan, Sept. 19, 1940, folder Home Guard, Records Group SG 12261, ASA, Montgomery.
29. Ida Lee Merchant, Secretary of the Central Trades Council, to Governor Dixon, Dec. 20, 1940, folder Home Guard, Records Group SG 12261, ASA, Montgomery.
30. Ibid.
31. Governor Dixon to Ida Lee Merchant of the Central Trades Council, Jan. 6, 1941, folder Home Guard, Records Group SG 12261, ASA, Montgomery.
32. Ibid.
33. The Disabled American Veterans of the World War in Russellville, Alabama, to Governor Dixon, Oct. 4, 1940, folder Home Guard, Records Group SG 12261, ASA, Montgomery.
34. Optimist Club of Montgomery to Governor Dixon, Dec. 13, 1940, folder Home Guard, Records Group SG 12261, ASA, Montgomery.
35. Telegram, Electrical Workers Union 833 of Jasper to Governor Dixon, Feb. 11, 1941, which referenced an article from the *Daily Press* of Jan. 14, 1941, folder Home Guard, Records Group SG 12261, ASA, Montgomery.
36. Ibid.
37. Reply, Governor Dixon to Electrical Workers Union 833, Feb. 16, 1941, folder Home Guard, Records Group SG 12261, ASA, Montgomery.
38. W. O. Rare, Secretary of the Alabama State Federation of Labor, to Governor Dixon, Feb. 17, 1941, folder Home Guard, Records Group SG 12261, ASA, Montgomery.
39. Ibid.
40. The Central Trades Council to Governor Dixon (n.d.), folder Home Guard, Records Group SG 12261, ASA, Montgomery.
41. State of Alabama, *Quadrennial Report*, p. 90.
42. Ibid., p. 92.
43. Ibid, p. 27.
44. Although induction of the National Guard had begun in September, 1940, the lack of available facilities for training forced the War Department to stagger the induction.
45. State of Alabama, *Quadrennial Report*, p. 76.
46. Changes in federal law had allowed the War Department to issue limited supplies of obsolete supplies and arms to state forces. However, federal forces and even allied forces came first, and the issue of federal equipment to state forces would not reach levels desired by state leaders until late in the war.
47. State of Alabama, *Quadrennial Report*, p. 76.
48. Ibid., p. 87.
49. Ibid.
50. Ibid., p. 76.
51. Ibid., p. 83–84.
52. Alabama State Military Department Office of the Adjutant General, *Annual*

Report of the Adjutant General of Alabama for the Fiscal Year Oct 1, 1945–Sep 30, 1946 (Montgomery: State of Alabama, 1946), p. 12–13.

53. State of Alabama, *Regulation Number 4*, p. 4.
54. Ibid., p. 7.
55. Subjects for training were suggested by Army Regulation 850-250 "Regulations for State Guards," paragraph 9. States usually used AR 850-250 as a base for their training plan, which was modified to fit local needs and resources.
56. "Proceedings of the Twenty-third Annual Convention of the American Legion, department of Alabama Huntsville, Alabama July 13, 14, 15, 1941," Mrs. Wilma G. Jacobs, stenotypist (bound transcript at the Headquarters of the American Legion, Montgomery, Ala.), p. 142.
57. Ibid., p. 35.
58. Ibid.
59. "Proceedings of the Twenty-Third Annual Convention," p. 69.
60. Ibid.
61. Ibid., pp. 70–71.
62. Ibid.
63. Ibid.
64. Ibid., pp. 88–89.
65. Ibid., p. 87.
66. Ibid., p. 88.
67. Colonel Cleere would become the state adjutant general from 1944 through 1946 and the end of the wartime State Guard.
68. Brig. Gen. Thomas J. Grayson, Adjutant General of the State of Mississippi, *Annual Report of the Adjutant General State of Mississippi from July 1, 1939 to June 30, 1941* (Jackson: State of Mississippi, 1941), pp. 5–6. Also, *ARCNGB Fiscal Year 30 June 1946* (Washington, D.C.: Government Printing Office, 1947), appendix 1.
69. Ibid., p. 7.
70. First authorized in the Mississippi Code, 1930. Paragraph 5485, Chap 136. Wartime State Guard organized by General Order Number 1, Feb., 19, 1941.
71. *Annual Report of the Chief National Guard Bureau 1941* (Washington, D.C.: Government Printing Office, 1941), p. 111.
72. The First Regiment, Mississippi State Guard, consisted of twelve lettered companies and a regimental headquarters company. On paper the force had 27 officers and 848 enlisted men (60 per company and 128 at headquarters). *Biennial Report of the Adjutant General of the State of Mississippi, from July 1, 1939 to June 30, 1941*, pp. 5–6.
73. Unlike most adjutants general, General Hays's rank came only from the state. He had no federal standing as a brigadier general.
74. Brig. Gen. Ralph Hays, acting adjutant general, *Biennial Report of the Adjutant General of the State of Mississippi from July 1, 1941, to June 30, 1943* (Jackson), p. 3.

75. General Hays described them in his report for the period of July, 1, 1941, to June, 30, 1943, as "professional men of all types, City and County officials and farmers who give freely of their services in order that our state may be adequately protected," p. 3.
76. Various 201 files of the State Guard officers of World War II and the Korean War, stored at the State Area Command, Jackson, Miss., headquarters of the Mississippi National Guard. Due to the private nature of the files, I was able to use them only on the condition that names would not be used.
77. Hays, *Biennial Report of the Adjutant General from July 1, 1941, to June 30, 1943*, p. 3.
78. Ibid., pp. 3–4.
79. Ibid.
80. State of Georgia to Governor Dixon, folder #2, Records Group SG 12242, Adjutant General, ASA, Montgomery.
81. Ibid.
82. Entry 226, folder 42494, box 472, RG 165, National Archives I, Washing-ton, D.C.
83. Editorial Staff of the War Records Committee, *Haverhill in World War II* (Haverhill, Mass.: City of Haverhill, 1946), p. 315.
84. Cyrille Leblanc and Col. Thomas F. Flynn, *Gardner in World War II* (Gardner, Mass.: Hatton Press, 1947), p. 25.
85. Rhode Island, which had created its State Guard of the Great War era through its Independent Chartered Companies, followed the more common state-directed approach to creating a State Guard in World War II. The Chartered Companies faded from state records in the late 1920s, and so Rhode Island recruited and outfitted a State Guard of one regiment in 1940 by Executive Order of Governor of October 25, 1940. *Annual Report of the Adjutant General of the State of Rhode Island for the Year 1941* (State of Rhode Island and Providence Plantations, n.d.), pp. 4–5.
86. William H. Murray, Ex-Governor of Oklahoma, "Memo For Posse Comitatus," folder Home Guard, Records Group SG 12261, ASA, Montgomery.
87. For a more complete explanation of the British Home Guard, see S. P. MacKenzie, *The Home Guard: A Military and Political History* (New York: Oxford University Press, 1995). Although MacKenzie takes no notice of similar forces in America, readers familiar with the American experience will find much at the community level that is similar.
88. Ibid., p. 124.
89. Ibid., pp. 82–83. This is MacKenzie's only inclusion of an American group coming to investigate the British Home Guard, and MacKenzie seemed more amused by the name of the committee than interested in its purpose.
90. Folder 324.5 "Home Guard File," box 470, RG 165, National Archives I, Washington, D.C. The War Department did not publish the report until its inclusion in the *Annual Report of the Chief of the National Guard Bureau* for 1946. Most likely this delay resulted not from security concerns but from the

lack of an *Annual Report* from 1943 through 1945. Appendix C to the *ARCNGB 1946.*

91. Files from the War Department contain hundreds of items gleaned from the British, for example, *Memorandum on the Feeding of the Home Guard* (1943), *Regulations for the Home Guard 1942* vols. 1 and 2, as well as transcripts of debates in Parliament on the Home Guard, "Information Circulars," "Instructions," and countless other items published by the British for use by their Home Guard. Folder 461 "General 20 British Home Guard Publications State Guard," Box 127, RG 168, National Archives II, College Park, Md.
92. Folder 353, box 120, RG 168, National Archives II, College Park, Md.
93. Folder 324, "State Guard 1941–1949," box 159, RG 168, National Archives II, College Park, Md. As more Regular soldiers were needed outside of the British Isles after the invasion threat passed, the Home Guard increasingly filled positions manning anti-aircraft guns and in coast artillery batteries.
94. J. K. Howard to MG William Bryden of the Office of the Chief of Staff, entry 226, folder 43166 "Equipment July 1, 1941–October 31, 1941," box 472, RG 165, National Archives I, Washington, D.C.
95. Folder "American Legion Plan," box 470, RG 165, National Archives I, Washington, D.C.
96. A confidential report dated October 31, 1941, lists twelve states (out of the forty-eight) that had not created State Guards. The states were Arizona, Arkansas, Idaho, Iowa, Louisiana, Montana, Nevada, North Dakota, Oklahoma, South Dakota, and West Virginia. Eventually, all of these states except for Arizona, Montana, Nevada, and Oklahoma created a State Guard. "Report to the Secretary of War on the Activities of the State Guards for the Year Ending October 31, 1941." RG 168, National Archives II, College Park, Md.

Chapter 6

1. In the Great War, a U.S. Army brigade contained four regiments. In the 1930s, the army adopted a new model for its division based on brigades of three regiments. The new, lighter "triangle" divisions were considered to be more mobile, while the older "squared" divisions with brigades of four regiments were deemed more suitable for defense. For this reason, the army delayed converting the National Guard divisions in the belief that in the next war the National Guard would play a defensive role rather than play a part in any expeditionary force.
2. Stetson Conn, *Guarding the United States and Its Outposts* (Washington, D.C.: Office of the Chief of Military History, Department of the Army, 1964), pp. 76–77.
3. Ibid.
4. Conn, *Guarding the United States,* pp. 76–77.
5. *ARCNGB 1946,* p. 41.

6. Memorandum for the Assistant Chief of Staff G-4, February 27, 1942, folder 43166 "January 1, 1942 THRU March 1, 1942" box 471, RG 165, National Archives I, Washington, D.C.
7. Folder 43166 "January 1, 1942 THRU March 1, 1942," box 47, RG 165, National Archives I, Washington D.C.
8. *ARCNGB 1946*, p. 42.
9. Folder 325.4, box 118, RG 168, National Archives II, College Park, Md.
10. Judge Advocate General, Jan. 5, 1942, folder "Opinions of JAG," box 470, RG 165, National Archives I, Washington, D.C.
11. Later in the war, about a dozen states began paying State Guardsmen for armory training, and most provided some pay for annual training, but unlike National Guardsmen, who were paid mostly by the federal government, State Guardsmen received all of their pay from their state government. See *ARCNGB 1946*, pp. 306–308.
12. Henry L. Stimson, Secretary of War, arguing against HR 6341, Jan. 14, 1942, folder 42494, box 470, RG 165, National Archives I, Washington, D.C.
13. Folder 324, box 117, RG 168, National Archives II, College Park, Md.
14. Maj. James Holland, New Hampshire State Guard, to General Marshall, Oct. 28, 1942, folder 324 "New Hampshire," RG 160, National Archives II, College Park, Md.
15. *History of the California State Guard* (n.p.: State of California, 1946), pp. 23–24.
16. H.R. 6397.
17. Folder 230.5, box 107, RG 168, National Archives II, College Park, Md.
18. Folder 324 "New Jersey," box 10, RG 160, National Archives II, College Park, Md.
19. The status of former State Guardsmen had only slight bearing on whether they would be commissioned or noncommissioned officers in the federal forces. Many enlisted State Guardsmen went to officers' candidate schools after completing basic training, while many State Guard officers became sergeants upon entering the federal forces.
20. "Table IX State Guardsmen Inducted Into Federal Service," from a report signed by Major General John F. Williams. Folder [torn] "Genera . . . State Guard," box 114, RG 168, National Archives II, College Park, Md.
21. U.S. Const. art. I, sec. 8.
22. Memorandum, General Williams to General John P. Smith, Chief of Administrative Services, Apr. 28, 1942, folder 350.001, box 118, RG 168, National Archives II, College Park, Md.
23. Folder 400.358, box 123, RG 168, National Archives II, College Park, Md.
24. Folder 324 "Massachusetts," box 10, RG 160, National Archives II, College Park, Md.
25. Transcript of telephone conversation, Colonel Camp of Georgia State Guard and Colonel Roemer, Aug. 22, 1942, folder SPAA 324.4, "Georgia," box 9, RG 160, National Archives II, College Park, Md.

26. *ARCNGB 1941*, pp. 22–23.
27. Adjutants General Association, *Annual Meeting The Adjutants General Association April 20–21, 1942, Washington DC* (n.p., Photostatic copy of typed transcript), p. 27.
28. Ibid., pp. 28–29.
29. Folder 324 "Maryland," box 10, RG 160, National Archives II, College Park, Md.
30. Ibid.
31. Ibid., p. 30.
32. Ibid., p. 31.
33. U.S. Const. art. I, sec. 8.
34. Adjutants General Association, *Annual Meeting The Adjutants General Association 1942*, p. 30.
35. Many states, such as Mississippi, used the term "State Guard Reserve" for an institution that is similar to today's Inactive National Guard. For them, the State Guard Reserve was a list of former State Guardsmen who did not attend drill but agreed to serve in an emergency.
36. An example of this type of militia group is the Reading (Mass.) Home Guard. Although no record of it exists in official town records, a few of the town's older residents remember patrolling utilities as part of the group. A few of the shoulder patches that the group created for itself survive as reminders of its existence.
37. *ARCNGB 1942*, p. 80.
38. Ibid., p. 83.
39. Folder "Opinions of JAG," box 470, RG 165, National Archives I, Washington, D.C.
40. Memorandum, Chief of Staff for the Secretary of War, Feb. 28, 1942, folder 34221 "STATUS,"box 472, RG 165, National Archives I, Washington, D.C.
41. Folder 342, box 118, RG 168, National Archives II, College Park, Md.
42. Red-bordered (immediate action) letter from Maj. Gen. John F. Williams, CNGB, Nov. 21, 1942, folder "Utah-State Guard-State file-unclassified," (*sic*), box 158, National Archives II, College Park, Md.
43. *Practical Home Guard Organization for Reserve Militia or "Minute Men"* (n.p.: National Rifle Association of America, 1942).
44. *Practical Home Guard Organization, Supplementary Bulletin Covering Application of the "Minute Man" to Large Cities* (n.p.: National Rifle Association of America, 1942).
45. Ibid., pp. 4–13.
46. *Practical Home Guard Organization for Reserve Militia or "Minute Men,"* p. 14.
47. Brig. Gen. S. Gardner Waller, *The Report of the Adjutant General of the State of Virginia Division of Military Affairs For the Period January 1, 1943, to December 31, 1943* (Richmond: Division of Purchase and Printing, 1944), p. 12.
48. Brig. Gen. S. Gardner Waller, *The Report of the Adjutant General of the State of Virginia Division of Military Affairs For the Period January 1, 1944, to*

December 31, 1944 (Richmond: Division of Purchase and Printing, 1945), p. 9.

49. John W. Listman, Jr., Robert K. Wright, Jr., and, Bruce D. Hardcastle, eds., *The Tradition Continues: A History of the Virginia National Guard, 1607–1985* (n.p.: Taylor Publishing Company, 1987), p. 187.
50. Adjutants General Association, *Annual Meeting The Adjutants General Association April 1–3, 1943, Harrisburg, PA* (n.p., Photostatic copy of typed transcript), pp. 78–79.
51. This request for federal support of state forces was based on Public Law 721, 77th Congress, approved Oct. 1, 1942.
52. *Annual Meeting The Adjutants General Association April 1–3, 1943*, p. 84.
53. Ibid., p. 90.
54. National Guard Association, *Annual Conference Report National Guard Association of the United States Organized 1878 [Including Addenda Executive Council Transactions 1940–1941–1942–1943] April 1,2 and 3, 1943, State Capitol Harrisburg Pa.* (n.p., Published transcript), pp. 24–25.
55. Ibid.
56. Ibid., p. 30.
57. Ibid., p. 32.
58. Ibid., p. 25.
59. Ibid., p. 32.
60. Ibid., p. 33.
61. *The State Defense Force Manual* (Harrisburg, Pa.: Military Service Publishing Company, 1941).
62. *Annual Conference Report National Guard Association 1940–April 1, 2, and 3, 1943*, p. 32.
63. *Annual Meeting The Adjutants General Association April 1–3, 1943*, p. 83.
64. For a fuller study of the relationship between the National Guard Association and the National Guard Bureau, as well as the lobbying power of the National Guard Association, see, Martha Derthick, *The National Guard in Politics* (Cambridge: Harvard University Press, 1965).
65. John K. Mahon, *History of the Militia and National Guard* (New York: Macmillan, 1983), p. 180.
66. The rating was A-I-J. *Annual Conference Report National Guard Association 1940–April 1, 2, and 3, 1943*, p. 32.
67. Adjutants General Association, *Special Meeting The Adjutants General Association June, 1943 Columbus, Ohio* (n.p., Photostatic copy of typed transcript), p. 58.
68. Ibid.
69. RG 165 box 472 folder 42494.
70. Letter to all Corps Area Commanders, Jan. 11, 1941, folder 421 "General 1 . . . 10 State Guard," box 125 RG 168, National Archives II, College Park, Md.
71. He had requested a rating of A-I-I instead of the A-I-J normally assigned to the State Guard.

72. Folder 324 "Virginia," box 10, RG 160, National Archives II, College Park, Md.
73. Folder 324 "Maryland," box 10, RG 160, National Archives II, College Park, Md.
74. Folder 370.02, box 121, RG 168, National Archives II, College Park, Md.
75. Adjutants General Association, *Special Meeting The Adjutants General Association June, 1943*, pp. 58–59.
76. Any request for federal pay for State Guardsmen would have been presumptuous, considering that members of the Organized Reserve (later the Army Reserve), would not receive pay for training until after World War II. For a further account of the decision to pay reservists for inactive training, see Richard B. Crossland and James T. Currie, *Twice the Citizen: A History of the Army Reserve, 1908–1983* (Washington, D.C.: Office of the Chief, Army Reserve, 1984), pp. 91–92.
77. Adjutants General Association, *Special Meeting The Adjutants General Association June, 1943*, pp. 60–61.
78. Ibid.

Chapter 7

1. C. A. Salisbury, *Soldiers of the Mists: Minutemen of the Alaska Frontier* (Missoula, Mont.: Pictorial Publishing Company, 1992), p. 57.
2. Folder SPAA 324.4 "Alaska," box 9, RG 160, National Archives II, College Park, Md.
3. Ronald H. Spector, *Eagle Against the Sun: The American War with Japan* (New York: Vintage Books, 1985), p. 178.
4. Salisbury, *Soldiers of the Mists*, p. 67.
5. Ibid., p. 73.
6. *Biennial Report of the Military Department State of Washington Office of the Adjutant General* (Olympia: State Printing Plant, 1943), pp. 3–6.
7. Ibid., pp. 6–8.
8. *Twenty-seventh Biennial Report of The Military Department of the State of Oregon to the Governor For the Period November 1, 1938, to October 31, 1940* (n.p.: State Printing Office, n.d.), p. 27.
9. *Twenty-eighth Biennial Report of The Military Department of the State of Oregon to the Governor For the Period November 1, 1940, to October 31, 1942* (n.p.: State Printing Office, n.d.), p. 2.
10. Ibid., p. 15.
11. Ibid., p. 3.
12. The total State Guard was organized into a headquarters, the First Regiment, a band, intelligence company, a military police platoon, an engineer battalion, and twenty-three separate battalions, nine troops of cavalry with two squadron headquarters.
13. *Twenty-eighth Biennial Report of The Military Department of the State of Oregon*, p. 22.
14. Ibid.

15. Office of the Adjutant General, State of California, *History of the California State Guard* (n.p.: State of California, 1946), p. 22.
16. Neither the Philippines nor Guam had a National Guard before World War II, and plans for their defense are well beyond the scope of the present study.
17. Charles Lamoreaux Warfield, "History of the Hawaii National Guard from Feudal Times to June 30, 1935" (M.A. thesis, University of Hawaii, 1932), p. 3.
18. I base these estimations on the rosters of Hawaii National Guard officers, from rosters of men on competition shooting teams, and of National Guardsmen receiving service bars for ten, fifteen, twenty, or twenty-five years of service. Colonel P. M. Smoot, *Annual Report of the Adjutant General Territory of Hawaii July 1, 1940, to June 30, 1941* (n.p.: Hawaii Printing Co., 1942), p. 20.
19. Gwenfread Allen, *Hawaii's War Years, 1941–1945* (Honolulu: University of Hawaii Press, 1950), p. 33.
20. Ibid.
21. Ibid., p. 38.
22. Allen, *Hawaii's War Years,* p. 149.
23. Ibid.
24. Smoot, *Annual Report of the Adjutant General Territory of Hawaii July 1, 1940, to June 30, 1941,* p. 3.
25. Allen, *Hawaii's War Years,* p. 149.
26. Ibid., p. 150.
27. Like the Organized Defense Volunteers, the British Home Guard existed to oppose an invasion rather than for civil defense, natural disaster, or domestic order. Also, the British Home Guard, like the Organized Defense Volunteers, formed a part of the national military forces. For a thorough study of the British experience with local defense during World War II, see MacKenzie, *Home Guard.*
28. Allen, *Hawaii's War Years,* p. 121.
29. Ibid., p. 95.
30. Ibid.
31. Ibid., p. 96.
32. Ibid., p. 97.
33. Ibid., p. 98.
34. William H. Conner and Leon de Valinger, Jr., *Delaware's Role in World War II, 1940–1946,* 2 vols. (Dover: Public Archives Commission, State of Delaware, 1955), 1:19.
35. Ibid.
36. Ibid., 2:97.
37. Ibid., 2:99.
38. Ibid.
39. Ibid., 2:100.
40. *South Dakota in World War II: An Account of the Various Activities of the People of South Dakota During the Period of World War II, both in South Dakota and*

Where South Dakotans and South Dakota Units Were Active Throughout the World (n.p.: World War II History Commission, n.d.), p. 604.

41. Ibid., p. 605. A fifth company was appointed a company commander, but the organization was stillborn and never mustered into state service because of an inability to recruit to strength.
42. Ibid., p. 604.
43. *Biennial Report of the Adjutant General of South Dakota July 1, 1940 to June 30, 1942* (Edward A. Beckwith, AG, n.p., n.d.). The federal government first issued .30 caliber Enfield rifles to the South Dakota State Guard. After the recall of 1942, the federal government replaced the Enfields with two Thompson submachine guns and thirty single-barrel twelve-gauge shotguns per company, p. 40.
44. Ibid., p. 42.
45. *South Dakota in World War II*, p. 604.
46. *Biennial Report of July 1, 1940 to June 30, 1942*, p. 41.
47. Ibid., p. 41.
48. Ibid.
49. List of "State Home Guard Plans." No date, but must be late 1940 or early 1941 at latest, folder 42494, box 472, RG 165, National Archives I, Washington D.C.
50. Ibid.
51. *ARCNGB 1941*, p. 112.
52. Folder 42494, box 472, RG 165, National Archives I, Washington, D.C.
53. Brig. Gen. Jay H. White, *State of Nevada Biennial Report of the Adjutant General for the Period July 1, 1940 to June 30, 1942, Inclusive* (Carson City: State Printing Office, 1942).
54. Brig. Gen. Jay H. White, *State of Nevada Biennial Report of the Adjutant General for the Period July 1, 1942 to June 30, 1944, Inclusive* (Carson City: State Printing Office, 1944), p. 5.
55. Ibid., pp. 9–10.
56. Ibid., p. 14.
57. Folder 42494, box 472, RG 165, National Archives I, Washington, D.C.
58. James E. Hancock, secretary, "[Biennial] Report of the Adjutant General State of Montana," Dec. 31, 1944. Office of the Adjutant General. Copy of typed report, at the National Guard Association library, Washington, D.C.
59. James E. Hancock, secretary, "[Biennial] Report of the Adjutant General State of Montana," Dec. 31, 1942. Office of the Adjutant General. Copy of typed report, at the National Guard Association library, Washington, D.C., p. 2.
60. Ibid.
61. Although War Department records contain documents denying authority for the Virgin Islands to create a home guard, shoulder patches exist from a "St. Thomas Home Guard" that existed from 1941 to 1943. Stephen D. Johnson

and Gary S. Poppleton, *Cloth Insignia of the U.S. State Guards and State Defense Forces* (Hendersonville, Tenn.: Richard W. Smith, 1993), p. 120.

62. D. F. Pancoast, Adjutant General, *Report of the Adjutant General of Ohio to the Governor of the State of Ohio for the Year 1941* (Columbus: F. J. Heer Printing, 1942), p. 3.
63. Ibid., p. 4. The adjutant general at the time doubled as the brigade commander.
64. Ibid., p. 6.
65. D.F. Pancoast, Adjutant General, *Report of the Adjutant General of Ohio to the Governor of the State of Ohio for the Year 1943* (Columbus: Heer Printing, 1944), p. 9.
66. *Report of the Adjutant General of West Virginia, 1940–1942* (Charleston, W.V.: Jarrett Printing Company, n.d.), p. iv.
67. Ibid., 224–27.
68. Government of Puerto Rico, Brig. Gen. Luis Raul Esteves, U.S. Army, Commanding, *Annual Report of the Puerto Rico State Guard (Confidential) for the Period Ending June 30, 1942* (San Juan: Bureau of Supplies, Printing, and Transportation, 1942), p. 5.
69. Unlike the Territories of Hawaii and Alaska, Puerto Rico opted to use the term "State Guard" ("Guardia Estatal" in Spanish) rather than "Territorial Guard" for its new militia.
70. Ibid.
71. Ibid., p. 6.
72. Puerto Rico, outside of the Anglo-American militia tradition shared by most of the states, had inherited an equally old and occasionally more thorough militia tradition from its Spanish past. For more information on the militia in Puerto Rico prior to the island's acquisition by the United States, see Luis A. de Casenave, *Historia de la Guardia Estatal de Puerto Rico, 1508–1992* (San Juan: Guardia Estatal de Puerto Rico, 1992).
73. Esteves, *Annual Report of the Puerto Rico State Guard for the Period Ending June 30, 1942*, p. 9.
74. Government of Puerto Rico, Brigadier General Luis Raul Esteves, U.S. Army, Commanding, *Annual Report of the Puerto Rico State Guard for the Period Ending June 30, 1943* (San Juan: Insular Procurement Office Printing Division, 1943), p. 10.
75. Esteves, *Annual Report of the Puerto Rico State Guard for the Period Ending June 30, 1942*, p. 10.
76. Esteves, *Annual Report of the Puerto Rico State Guard for the Period Ending June 30, 1943*, p. 5.
77. Ibid., p. 13.
78. *Report of the Adjutant General of the State of Connecticut to the Governor and the Assembly for the Period Jan. 1, 1940 to Dec. 31, 1940* (n.p., n.d.), p. 25.
79. Ibid.

80. Ibid., p. 29.
81. Ibid., p. 28.
82. *Report of the Adjutant General of the State of Connecticut to the Governor and the General Assembly for the period Jan. 1, 1941 to Dec. 31, 1941* (n.p., n.d.), p. 10.
83. Mrs. Eleanor Grant Rigby, "Report of the Commandant of the Motor Corps," (pp. 11 and 12 of the *Report of the Adjutant General of the State of Connecticut* for 1941).
84. *Report for the period Jan. 1, 1941 to Dec. 31, 1941*, p. 12.
85. The restructuring of the army in 1942 that had produced the Army Ground Forces, Army Air Forces, and the Army Service Forces had replaced the earlier system of Corps Areas with a new system by which the U.S. Army divided the continental United States into Service Commands, which were part of the Army Service Forces.
86. *Boston Herald,* June 23, 1943, p. 100.
87. *Christian Science Monitor,* June 17, 1942, p. 79.
88. Ibid., p. 78.
89. *Boston Herald,* June 18, 1942, p. 77.
90. *Boston Herald,* Aug. 7, 1942, p. 81.
91. *Christian Science Monitor,* June 17, 1942, p. 79.
92. Brig. Gen. Raymond H. Flemming, AG, *Report of the Adjutant General of the State of Louisiana January 1, 1940, to December 31, 1941 to the Governor and Legislature of the State of Louisiana* (n.p., n.d.), p. 24.
93. Brig. Gen. Raymond H. Flemming, AG, *Report of the Adjutant General of the State of Louisiana January 1, 1941, to December 31, 1942 to the Governor and Legislature of the State of Louisiana* (n.p., n.d.), p. 13.
94. Ibid., pp. 42–43.

Chapter 8

1. Letter, Jan. 29, 1943, folder 200.4, box 106, RG 168, National Archives II, College Park, Md.
2. An example of this is an obituary in the North Reading (Mass.) *Transcript* of June 5, 1997. The obituary for a George P. Luther, who died at the age of ninety-two, mentions that "during World War II he was an electrician at the Charlestown Navy Yard and served in the National Guard." As Luther would have been thirty-six or thirty-seven years old at the time of the entry of the United States into the war, and the obituary gives no indication of any other military service, Luther in all probability served in the Massachusetts State Guard and not the National Guard during the war.
3. Some 1,773 officers, 43 warrant officers, and 94,227 enlisted men were discharged during this period. *ARCNGB 1941*, p. 18.
4. Maj. Gen. John Williams, folder 210.1, box 106, RG 168, National Archives II, College Park, Md.

5. Maj. Gen. Sherman Miles, Mar. 20, 1943, folder "Vermont—State Guard—State File-Classified 1941–1943," box 158, RG 168, National Archives II, College Park, Md.
6. Maj. Gen. W. D. Styer to Maj. Gen. Sherman Miles, Mar. 24, 1943, folder "Vermont—State Guard—State File-Classified 1941–1943," box 158, RG 168, National Archives II, College Park, Md.
7. Folder "Defense Plan State of Alabama (Defense Plan #1, 1 October 1940)," box 158, RG 168, National Archives II, College Park, Md.
8. Folder SPAA 324.4 "Alaska," box 9, RG 160, National Archives II, College Park, Md.
9. Circular letter, CNGB, Apr. 8, 1943, folder "Pennsylvania State Guard—State File 1941–1944," box 157, RG 168, National Archives II, College Park, Md.
10. *ARCNGB 1946*, p. 248, "Appendix G." States lost officers at a lesser rate because State Guard officers tended, especially in the early years, to be too old for conscription, many being former Great War officers.
11. Folder 000.7, box 104, RG 168, National Archives II, College Park, Md.
12. Lt. Col. E. T. Kimball, of the NGB, to Illinois Military Department, folder SPAA 324.4 "Illinois," box 9, RG 160, National Archives II, College Park, Md.
13. Folder 011, box 104, RG 168, National Archives II, College Park, Md.
14. Brig. Gen. S. Gardner Waller, Adjutant General, *The Report of the Adjutant General of the State of Virginia Division of Military Affairs For the Period January 1, 1944, to December 31, 1944* (Richmond: Division of Purchase and Printing, 1945), p. 10.
15. D. F. Pancoast, *Report of the Adjutant General of Ohio to the Governor of the State of Ohio for the Year 1943* (Columbus: Heer Printing, 1944), p. 4.
16. Col. L. M. Hart, report to the Chief National Guard Bureau, Aug. 28, 1944, folder "Maine State Guard-state file 1941–1945," box 135, RG 168, National Archives II, College Park, Md.
17. Western Defense Command Counter Fifth Column Plan, 1944, at the Army War College Library, Carlisle Barracks, Pa.
18. BG S. Gardner Waller, *The Report of the Adjutant General of the State of Virginia Division of Military Affairs For the Period January 1, 1943, to December 31, 1943* (Richmond: Division of Purchase and Printing, 1944), p. 12.
19. Curren R. McLane, "A History of the Texas State Guard," Headquarters, Texas State Guard, [1983?], pp. 12–13.
20. John Morton Blum, *V Was for Victory: Politics and American Culture During World War II* (New York: Harcourt Brace Jovanovich, 1976), p. 202.
21. Brig. Gen. Le Roy Pearson *Report of the Adjutant General of Michigan for the Period January 1, 1943 to June 30, 1944* (n.p., [1944]), p. 6.
22. Alfred McClung Lee and Norman D. Humphrey, *Race Riot: Detroit, 1943* (New York: Octagon Books, 1968), p. 39.
23. Pearson, *Report of the Adjutant General,* p. 8.
24. Lee and Humphrey, *Race Riot,* pp. 52–53.
25. Pearson, *Report of the Adjutant General,* p. 9.

26. Folder "Pennsylvania State Guard—State File 1941–1944," box 157, RG 168, National Archives II, College Park, Md.
27. Brig. Gen. Walter J. Delong, document, Sept. 10, 1944, folder "Washington State Guard file 1941–1944," box 158, RG 168, National Archives II, College Park, Md.
28. *ARCNGB 1946*, p. 42.
29. Service Commanders to General Somervell, Feb. 26, 1944, folder 413.44, box 124, RG 168, National Archives II, College Park, Md.
30. The Adjutants General Association, *Annual Meeting The Adjutants General Association May 3–4–5–6, 1944 Baltimore, Maryland* (n.p., Photostatic copy of typed transcript), p. 29.
31. Ibid., p. 29.
32. National Guard Association, *Annual Conference Report National Guard Association of the United States Organized 1878 Sixty-Sixth Annual Convention at The Hotel Southern May 3–4–5–6 1944 Baltimore Maryland* (n.p., published transcript), pp. 198–99.
33. Folder 312.3, box 107, RG 168, National Archives II, College Park, Md.
34. Folder 451.2 "General State Guard," box 126, RG 168, National Archives II, College Park, Md.
35. Ibid., p. 73.
36. Brig. Gen. Ralph Hays, *Biennial Report of the Adjutant General of the State of Mississippi from July 1, 1943–June 30, 1945* (Jackson: n.p., n.d.), 6–8.
37. Ibid.
38. Hays, *Adjutant General's Report of July 1, 1941–June 30, 1943*, p. 4.
39. Ibid., p. 8.
40. "Report by Brig. Gen. William P. Wilson, Adjutant General to the members of the legislature, State of Mississippi, for the biennial period ending 30 June, 1947, part II." In Records Group 33, vol 23, at the Mississippi State Archives, Jackson. The previous report placed attendance figures for armory drill at 80 percent (p. 7).
41. Ibid., p. 1.
42. The heavy weapons platoon was equipped with Browning machine guns, which each took a four-man crew.
43. W. B. Alexander to Senator James O. Eastland, Dec. 14, 1944, folder 324, box 117, RG 168, National Archives II, College Park, Md.
44. *Daily Clarion-Ledger* (Jackson), Mar. 30, 1944, p. 1.
45. *Annual Report of the Chief National Guard Bureau, Fiscal Year 30 June 1946* (Washington, D.C.: Government Printing Office, 1947), appendix I. As the National Guard Bureau did not publish an annual report for the years 1943 through 1945, the 1946 report included the data for the missing years.
46. *Jackson Daily News*, Apr. 11, 1944, p. 3. The total number of men activated in response to the labor dispute came to 33 officers and 348 enlisted men, making it the largest mobilization of the Mississippi State Guard.
47. *Clarksdale Daily Press*, June 23, 24, 1944, p. 1.

48. Ibid., Jan. 15, 1945, p. 1.
49. *Annual Report of the Chief National Guards Bureau, Fiscal Year Ending 30 June 1946*, appendix J, table III, also, *Daily Clarion Ledger* (Jackson), Dec. 4, 1945, p. 1.; Oct. 8, 1946, p. 1.
50. From its organization until June 30, 1946, the Mississippi State Guard lost 46 officers and 1,243 enlisted men through induction into the federal armed forces. From *Report of the Chief National Guard Bureau Fiscal Year Ending 30 June 1946*, appendix G.
51. Hays, *Biennial Report for July 1, 1943 to June 30, 1945*, pp. 6–7. General Hays based his assertion that prior State Guard service benefited selectees on letters his department received from selectees after they had entered the federal military, who sought proof of their state service as a means to attain NCO status.
52. Mississippi House Bill No. 366, 1944.
53. General Wilson, "Report to the Members of the Legislature for the biennium ending 30 June, 1947." Part III.
54. Ibid.
55. Ibid.
56. "Annual Meeting The Adjutants General Association Hotel Fort Des Moines 1945 Des Moines, Iowa" (photostatic copy of typed transcript), at NGA Library, Washington, D.C., pp. 12–16.
57. *Report of the Chief National Guard Bureau Fiscal Year Ending 30 June 1946*, pp 88–89, fn. 9.
58. The Mississippi State Guard Reserve was similar to the State Guard Reserve in Oregon, that is, a list of inactive State Guardsmen that could be called upon to form units if needed. No record exists of Mississippi having "Minute Man" or other "Reserve" type units in its State Guard.
59. Mississippi Special Orders 224, "officers promoted one rank and transferred to the State Guard Reserve (inactive) in accordance with 'section 8541, code of Mississippi, 1942' and with customs of the service." Orders found in various 201 files of World War II State Guard officers at the MS-STARC.
60. Adjutants General Association, *Annual Meeting The Adjutants General Association Hotel Statler 1946 Washington, D.C.* (n.p., Photostatic copy of typed transcript), pp. 73–74.
61. Ibid., pp. 112–13.
62. Letter, Apr. 25, 1944, folder 324, box 117, RG 168, National Archives II, College Park, Md.
63. The official name of any New York militia that was not National Guard or naval militia has long been the "New York Guard." The term "State Guard" was used by the state in a general sense, but the official name was the "New York Guard."
64. The Hugh A. Drum Papers, folder "Retirement 1943," box 26, U.S. Army Military History Institute, Carlisle Barracks, Pa.
65. Ibid., folder "Clippings on Drum's Appointment as NY Guard Commander."

66. Ibid., folder "Requests for Commissions."
67. Ibid., letter, Mar. 21, 1944, folder "New York State Guard."
68. Ibid., folder "Legislation Re Pay For Off. & E.M. AT Armory Drill 1/31/45."
69. Ibid., folder *"Confidential."*
70. Ibid., folder "NY Internal Security Plan."
71. Ibid., "Circular Number 8," Draft 1, Nov., 1946, *Re-establishment of the NYNG.*
72. Ibid., folder "New York Guard Emergency Plan," box 27.
73. Adjutant General of Indiana to the War Department, Mar., 1946, folder 413.3, box 124, RG 168, National Archives II, College Park, Md.
74. BG Charles H. Grahl, *Biennial Report of the Adjutant General of the State of Iowa for the Fiscal Years 1945 and 1946 Beginning July 1, 1944 and ending June 30, 1946* (Des Moines: The State of Iowa, 1944), p. 7.
75. The Virginia State Guard Reserve had been disbanded following the end of the war.
76. Article VI—Unorganized Militia—The Military Code of Virginia, including Sections 1, 2, 3, 4, and 5.
78. Brig. Gen. S. Gardner Waller, Adjutant General, *The Report of the Adjutant General of the State of Virginia Division of Military Affairs For the Period January 1, 1946, to December 31, 1946* (Richmond: Commonwealth of Virginia Division of Purchase and Printing, 1948), p. 9.
78. Ibid., p. 9.
79. Ibid., p. 12, "General Orders No. 8," The Adjutant General's Office.
80. Brig. Gen. S. Gardner Waller, Adjutant General, *The Report of the Adjutant General of the State of Virginia Division of Military Affairs For the Period January 1, 1947, to December 31, 1947* (Richmond: Commonwealth of Virginia Division of Purchase and Printing, 1949), p. 15.
81. William H. Conner and Leon de Valinger, Jr., *Delaware's Role in World War II, 1940–1946*, 2 vols. (Dover: Public Archives Commission, State of Delaware, 1955), 2:100.
82. Ibid., 2:149.
83. Folder 200.6, box 106, RG 168, National Archives II, College Park, Md.
84. George A. Craig, *History: Massachusetts State Guard and Massachusetts State Guard Veterans* (Boston: Spaulding-Moss, 1931).
85. Capt. George A. Craig, who had served in the Massachusetts State Guard of the Great War and who later served during the Boston Police Strike of 1919, May 25, 1947, folder 011, box 104, RG 168, National Archives II, College Park, Md.

Chapter 9

1. Mahon, *History of the Militia and the National Guard*, pp. 201–2.
2. Brig. Gen. William P. Wilson, *Biennial Report of the Adjutant General of the State of Mississippi to the Governor and Legislature for the Biennium ending June 30, 1949* (n.p.: Jackson, n.d.), 17.

3. George J. Stein, "State Defense Forces: The Missing Link in National Security," *Military Review* 64 (Sept., 1984): 2–16.
4. Army Regulation 915-10, *State Guards, General Policy and Regulation for State Guards,* 1950.
5. Derthick, *National Guard in Politic,* p. 112.
6. *ARCNGB 1951,* p. 1.
7. Public Law 849, 81st Congress, Dec. 27, 1950.
8. "Official Proceedings of the National Guard Association of the United States 74th General Conference at the Hotel Shirley-Savoy Denver, Colorado, 6–9 October 1952." At the National Guard Association [hereafter cited as the "NGA"] Library, Washington, D.C.
9. Maj. Gen. C. D. O'Sullivan, *Biennial Report of the Adjutant General July 1, 1946 to June 30, 1948* (n.p., n.d.), p. 39.
10. Ibid., "Circular No. 3," published on Feb. 17, 1947. 11. Maj. Gen. C. D. O'Sullivan, *Biennial Report of the Adjutant General July 1, 1948 to June 30, 1950* (n.p., n.d.), p. 64.
12. "Circular No. 33," State of California Office of the Adjutant General, Dec. 8, 1948.
13. O'Sullivan, *Biennial Report of the Adjutant General July 1, 1948 to June 30, 1950,* p. 64.
14. "Statutes 1949, Chapter 678," which added Chapter 3, to Part 2, Division 2, of the California Military and Veterans Code.
15. "Circular No. 42," from the Office of the Adjutant General, Oct. 10, 1949.
16. O'Sullivan, *Biennial Report of the Adjutant General July 1, 1948 to June 30, 1950,* p. 64.
17. Ibid.
18. Ibid., p. 65.
19. Maj. Gen. Earle M. Jones, *Biennial Report of the Adjutant General For the Period 1 July 1950 to 30 June 1952* (n.p., n.d.), p. 112.
20. "Section 4088," 81st Congress, amended "Section 61A" of the *National Defense Act* to allow states to create state military forces not in the National Guard.
21. From sergeant first-class through general, the number of slots for each rank were the same at full or cadre strength. The difference in numbers came at the lower ranks. While a fully manned division had 803 sergeants, a cadre had 752. From there the differences became more pronounced. For corporal the numbers were 870 versus 236, private first-class were 2,326 versus 25. As expected, the biggest difference came at the lowest rank, private. Whereas a fully manned division contained 1,992, almost a third of its strength, a cadre division had none. The TOE contained no rank of staff sergeant for Internal Security forces.
22. Jones, *Biennial Report of the Adjutant General For the Period 1 July 1950 to 30 June 1952,* p. 113.
23. Ibid., p. 114.

24. *Thirty-second Biennial Report of the Military Department of the State of Oregon to the Governor for the Period January 1, 1949 to December 31, 1950* (n.p.: State Printing Dept., n.d.), p. 6.
25. The cadre concept for State Guards during the Korean War was followed by Connecticut, Hawaii, Indiana, Ohio, Oregon, Rhode Island, and Washington.
26. *Thirty-third Biennial report of the Military Department of the State of Oregon to the Governor for the Period January 1, 1951 to December 31, 1952* (n.p.: State Printing Dept, n.d.), p. 6.
27. Ibid., p. 7.
28. Valentine J. Belfiglio, *Honor, Pride, Duty: A History of the Texas State Guard* (Austin: Eakin Press, 1995), p. 66.
29. Maj. Gen. Kearie L. Berry, *Annual Report of the Adjutant General for the Fiscal Year Ending August 31, 1951* (Austin: Adjutant General's Dept., 1950), p. 1.
30. N.Y. Const. art. 12, para. 3.
31. State of New York, *Annual Report of the Chief of Staff to the Governor For the Division of Military and Naval Affairs for the Year 1950* (Albany: Williams Press, 1951), p. 117.
32. "Public Law 849," 81st Congress.
33. *ARCNGB 1951*, pp. 20–22.
34. State of New York, *Annual Report of the Chief of Staff to the Governor For the Division of Military and Naval Affairs for the Year 1951* (Albany: Williams Press, 1951), p. 121.
35. Ibid., p. 122.
36. Unlike Pennsylvania, Massachusetts, New Jersey, and Vermont, which had only planned a State Guard, Connecticut actually created a cadre force during the Korean War.
37. U.S. Const. art. I, sec. 10.
38. S. 968, and, H.R. 3210.
39. "Public Law 435."
40. S 2984, introduced by Senator Ives, and, HR 7445, introduced by Representative Radwan, both on Apr. 8, 1952.
41. "Official Proceedings of the National Guard Association of the United States at Hotel Shirley-Savoy, Denver, Colorado, 6–9 October, 1952." At the NGA Library, Washington, D.C.
42. State of New York, *Annual Report of the Chief of Staff to the Governor For the Division of Military and Naval Affairs for the Year 1954* (New York: Herald Square Press, 1955), p. 8.
43. Brig. Gen. William P. Wilson, *Biennial Report of the Adjutant General of the State of Mississippi for the Biennium Ending 30 June 1953* (Jackson: n.p., n.d.), 8.
44. The years just prior to the outbreak of the Korean War show few state activations of the Mississippi National Guard. The *Biennial Report of the Adjutant General of Mississippi for the Period Ending 30 June 1949* lists an August, 1947, raid in Rankin County for alcohol, a hurricane in September, 1947, two

floods in February, 1948, and tornados in February, 1948, and January and March, 1949 (pp. 5–6).

45. "Circular Letter," May 11, 1951, from the Acting Chief, NGB, to all state Adjutants General. At the NGA Library, Washington, D.C.
46. Wilson, *Biennial Report of the Adjutant General of the State of Mississippi for the Biennium Ending 30 June 1953*, p. 9.
47. Ibid, p. 26.
48. From the 201 files, individual service records, of State Guard officers from World War II and the Korean War at the MS-STARC.
49. "Special Order number 64," Apr. 3, 1951. "Special Order number 62," Mar. 31, 1951, commissioned the four officers in the State Guard Reserve. In officer's 201 file at the MS-STARC.
50. *Daily Clarion-Ledger* (Jackson), Apr. 1, 1951. "Holmes Gambling Joints Raided. Guardsmen Carry Warrants for Twenty Places in Cleanup. Thirty Arrested by State Guard." p. 1.
51. "Special Orders number 98," May 19, 1951, in officer's 201 file at the MS-STARC, also *Daily Clarion-Ledger* (Jackson), May 20, 1951, p. 1.
52. Brig. Gen. John O'Keefe, *Biennial Report of the Adjutant General of the State of Mississippi from July 1, 1937 to June 30, 1939* (Jackson: n.p., n.d.), 6.
53. "Special Orders Number 97," Apr. 30, 1952, in officer's 201 file at the MS-STARC, also *Daily Clarion-Ledger* (Jackson), Apr. 24, 1952, p. 1.
54. Wilson, *Biennial Report of the Adjutant General for the Biennium Ending June 30, 1953*, pp. 24–25.
55. *Daily Clarion-Ledger* (Jackson), Apr. 1, 1951, p. 1, also "Special Orders Number 62," Mar. 31, 1951, in officer's 201 file, at the MS-STARC.
56. Application for Commission in the Mississippi State Guard Reserve, officer's 201 file, at the MS-STARC.
57. States that, along with Florida, took this tentative planning approach to creating a State Guard during the Korean War were Arkansas, Georgia, Kentucky, Louisiana, Maryland, New Jersey, South Carolina, and Vermont. In addition, Illinois, Massachusetts, and Pennsylvania made plans for large forces but never got beyond the planning stage.
58. "Chapter 251 Florida Statutes," in Mark W. Lance, Maj. Gen., *Report of the Adjutant General of the State of Florida for the Years 1951 and 1952* (n.p., [1952]), p. 38.
59. Ibid., p. 39.
60. Throughout the reports of the period, the force is referred to as "Michigan State Troops (State Guard)" formally, and simply as "State Troops" otherwise.
61. Military Section, Adjutant General's Office, *Biennial Report of the Adjutant General of Michigan For the Period 1 July 1948 Through 30 June 1950 (with Addenda to 15 May 1951)* (n.p., [1951]), p. 19.
62. These were authorized under Michigan "Public Law 849."
63. *Report of the Adjutant General [of Michigan] 1952*, Mar. 16, 1953 (n.p., n.d.), p. 37.

64. William J. Watt and James R. H. Spears, eds., *Indiana's Citizen Soldiers: The Militia and National Guard in Indiana History* (Indianapolis: Indiana State Armory Board, 1980.), p. 181.
65. Title 32, sec. 109, U.S. Code.
66. Ibid., p. 182.
67. Lt. Col. Merrill H. Huntzinger, IGR, "2nd Brigade, Management of the Greenfield Armory," in Col. Felix E. Goodson, *The Indiana Guard Reserve* (Indianapolis: Indiana Creative Arts, 1998), pp. 130–32.
68. Maj. Gen. George M. Haskett, AG, *Biennial Report of the Adjutant General of Washington July 1, 1960 to June 30, 1962* (n.p.: State of Washington Military Department, 1962), [not paginated].
69. Ibid.
70. Maj. Gen. Howard S. McGee, AG, *Biennial Report of the Adjutant General July 1, 1972–June 30, 1974* (n.p.: State of Washington Military Department, 1975), [not paginated].
71. North Dakota "Executive Order #30," Jan. 25, 1966.
72. Maj. Gen. La Clair A. Melhouse, *Biennial Report 1 July 1964–30 June 1966* (n.p.: Adjutant General's Department, State of North Dakota, 1966), Annex A, A-2.
73. Douglas R. Hartman, *Nebraska's Militia: The History of the Army and Air National Guard, 1854–1991* (n.p.: Donning Company/Publishers, 1994), p. 135.
74. Mahon, *History of the Militia and the National Guard*, p. 234.
75. Ibid., p. 235.
76. Hartman, *Nebraska's Militia*, p. 199.
77. Nebraska Military Department, *Biennial Report Fifty-Seventh Biennial Report Period 1 July 1966–30 June 1968* (n.p.: Nebraska Military Department, n.d.), p. 8.
78. Ibid.
79. Hartman, *Nebraska's Militia*, p. 199.
80. *Annual Report Department of the Adjutant General State of Maine 1 July 1969–30 June 1970* (n.p., 1970), pp. 76–77. The Maine State Guard was organized into one brigade, which had five battalions (the First, Second, and Third companies of the 103rd Battalion, and the First and Second of the 314th). In addition, the force had headquarters companies for the brigade and the battalions, p. 68.
81. Ibid., p. 76.
82. Ibid., p. 78.
83. Ibid., p. 76. Authorized June 28, 1965.
84. Ibid., p. 75.

Chapter 10

1. Lewis Sorley, "Creighton Abrams and Active-Reserve Integration in Wartime," *Parameters* 21 (Summer, 1991): 35–50, see also, Lewis Sorley, *Thunderbolt from the Battle of the Bulge to Vietnam and Beyond: General Creighton*

Abrams and the Army of His Times (New York: Simon and Schuster, 1992), pp. 361–64.

2. Exact numbers are not possible because of the nature of these forces. Some units existed only on paper, while others were formed at a cadre level only to fade away after a few years.
3. Brig. Gen. Bruce Jacobs, "Memorandum For the Record," June 16, 1980, at the National Guard Association [hereafter cited as "NGA"] Library, Washington, D.C.
4. Grace P. Hayes, Vice President of HERO, to the Adjutants General of all states, Aug. 21, 1980, at the NGA Library, Washington, D.C.
5. HERO, *U.S. Home Defense Forces Study* (Historical Research and Evaluation Study, Prepared for the Office of the Assistant Secretary of Defense under Contract, 1981).
6. "Memorandum for: General Jacobs, Colonels Allen and Perkins," July 14, 1981, at the NGA Library, Washington, D.C. The memorandum, apparently an internal memo of the NGA, suggested that General Coates also be used to investigate any potential involvement of the NGA with the State Guard.
7. "Comments by MG George E. Coates, TAG, WA," read at the Adjutants General Conference, Nashville Tenn., Jan. 19, 1982, at the NGA Library, Washington, D.C.
8. Ibid.
9. Headquarters, Department of the Army, Draft of "State Defense Forces: Policy and Guidance for State Defense Forces," Mar. 2, 1982, at the NGA Library, Washington, D.C. Later, the army would amend 670-1, its regulation regarding uniforms, to require that members of purely state forces adopt a red name plate on their dress uniforms, to further differentiate state soldiers from federal soldiers, who wore a black name plate.
10. From draft of slides for briefing of Mar. (May?) 17, 1983, to General Wickham on State Defense Forces, at the NGA Library, Washington, D.C.
11. Roger A. Beaumont, "Constabulary or Fire Brigade?: The Army National Guard," *Parameters: Journal of the U.S. Army War College* 12 (Spring, 1982): 62–69.
12. Bruce Jacobs, "Memo for Record: State Military Forces after the Guard Is Called," *National Guard* 36 (June, 1982): 40.
13. "B.J." [Brig. Gen. Bruce Jacobs?], "Memorandum for the Record" Aug. 19, 1982, at the NGA Library, Washington, D.C.
14. Lt. Gen. Emmett H. Walker, Jr., Chief, NGB, to the Adjutants General of all States, Puerto Rico, the Virgin Islands, Guam, and the District of Columbia, Nov. 19, 1983, at the NGA Library, Washington, D.C.
15. Proposal "Relating to the Establishment of Federal Support for State Defense Forces," at the 104th General Conference of the National Guard Association, Sept. 22, 1982, at the NGA Library, Washington, D.C.
16. "Information Paper Subject: Support of State Defense Forces," Sept. 8, 1982. Attached to above, at the NGA Library, Washington, D.C.

17. U.S. Const. art. I, sec 8.
18. George Kearney, ed., *Official Opinions of The Attorneys General of the United States Advising the President and Heads of Departments in Relation to Their Official Duties* (Washington, D.C.: Government Printing Office, 1929), 29:322–29. This ruling had prevented the National Guard from crossing into Mexico and had led to the dual oath and enlistments for the National Guard in the *National Defense Act of 1916*. See chapter 1.
19. Nolan E. Jones to Maj. Gen. Bruce Jacobs, Dec. 8, 1982, at the NGA Library, Washington, D.C.
20. Maj. Gen. Eston Marchant, Adjutant General, *Report of the Adjutant General of South Carolina fiscal Year 1 July 1981–30 June 1982* (n.p.: Printed Under the Direction of the State Budget and Control Board, n.d.), pp. 189–90. New SDFs would also be formed in Alabama, Georgia, Maryland, and Michigan.
21. John A. Crosscope, Jr., "Position Paper—Draft Regulation, State Defense Forces," June 9, 1982, at the NGA Library, Washington, D.C.
22. Ibid.
23. Jerry Fogel of Martin Fromm and Associates, to all attendees, Nov. 18, 1983, at the NGA Library, Washington, D.C.
24. From undated sheet entitled "Who Is Jerry Fogel," at the NGA Library, Washington, D.C.
25. Telephone interview with Jerry Fogel, Oct. 4, 2000. Martin Fromm was Jerry Fogel's father-in-law.
26. Memo, Jerry Fogel to the Steering Committee of State Defense Association of the U.S., Nov. 18, 1983, at the NGA Library, Washington, D.C. Also sought were a former governor of Tennessee and another of North Carolina, a congressman, the head of the Association of the U.S. Army, and one other man whose title was not listed.
27. Ibid.
28. Jerry Fogel "Memo to: Steering Committee State Defense Association of the U.S.," Dec. 15, 1983, at the NGA Library, Washington, D.C.
29. Jerry Fogel "Memo to: Steering Committee State Defense Association of the U.S.," Dec. 16, 1983, at the NGA Library, Washington, D.C..
30. Brig. Gen. John A. Crosscope, Jr., to Jerry Fogel, Dec. 11, 1984, at the NGA Library, Washington, D.C.
31. Brig. Gen. John A. Crosscope, Jr., to Lt. Gen. LaVern E. Weber, Executive Director, National Guard Association, Dec. 12, 1984; and, Brig. Gen. John A. Crosscope, Jr., to Lt. Gen. Emmett H. Walker, Jr., Chief, National Guard Bureau, Dec. 12, 1984, at the NGA Library, Washington, D.C.
32. George J. Stein, "State Defense Forces: The Missing Link in National Security," *Military Review* 64 (Sept. 1984): 2–16.
33. Ibid., pp. 13–15.
34. Brig. Gen. John A. Crosscope, Jr., to Jerry Fogel, Dec. 11, 1984, at the NGA Library, Washington, D.C. Most of Stein's information on home guard forces had come from the HERO study.

35. Fogel had joined an organization called the Missouri Reserve Force and began signing himself as "Major, MORF" on SDFAUS correspondence. He broke all connections with the Missouri Reserve Force when he discovered that it was not a state-sanctioned force but a private organization. From telephone interview with Jerry Fogel, Oct. 4, 2000.
36. *SDFAUS News* 1, no. 1 (May/June 1985).
37. By the summer 1986, Indiana, Louisiana, and Maryland would also open state chapters of the SDF Association. However, the SDFAUS would also count thirty-two members from Massachusetts, which had no state chapter, and another five members each from Oregon and Virginia, three from Alabama, two from Nevada, and one each from Maine and Montana. Montana never created a State Guard in the twentieth century.
38. Ibid., "From Headquarters." Early application brochures for the SDFAUS featured a photo of Gov. George Deukmejian of California receiving the SDFAUS's Man of the Year award, as well as a photo of Fogel shaking hands with Vice President George Bush.
39. In 1985, Fogel would be appointed by Missouri governor John Ashcroft to the Missouri Commission on Crime. The forty-one-member commission, chaired by the state's attorney general, was created, among other tasks, "to enhance Missouri's ability to monitor paramilitary groups." From a news release of Aug. 14, 1985, of Martin Fromm and Associates. As late as October, 1986, Fogel would still be writing letters to the NGB as executive director of the SDFAUS.
40. Robert Herzberg, "Washington `Hot Line,'" *Combat Arms,* Mar., 1985, p. 12.
41. Michael Pietrantoni, "True Militia: Citizen Defense Forces," *American Survival Guard,* Oct., 1985, p. 32.
42. Armond Noble, "State Defense Forces—The New `Third Army,'" *Military* 2 (Feb., 1986): 11–12. Noble was the publisher of *Military* as well as a warrant officer in the California State Military Reserve.
43. "The Few, The Pointless: The Texas State Guard," article from undetermined source [*Reporter?*], probably from mid-1980s, at the NGA Library, Washington, D.C.
44. "Disbanded Guard Unit Regrouping," *Dallas Morning News,* sec. A, p. 1, 14. Undated newspaper clipping located at the NGA Library, Washington, D.C.
45. James L. Pate, "Citizen Soldiers: Fighting for the Right to Defend America," *Soldier of Fortune,* May, 1978, p. 59. The story of Major Holloway featured in a magazine commonly geared toward mercenaries and survivalists did little to win supporters for SDFs within the army, the National Guard, state governments, or the mainstream media.
46. "Texas State Guard May Have to Fight for Fiscal Survival," *Houston Post,* Apr. 28, 1985, sec. D, p. 7.
47. Article 5786, sec 8, Texas Code.
48. Pate, "Citizen Soldiers," pp. 59, 61, 85.
49. Pate, "Citizen Soldiers," p. 86. The name Holloway chose for the corporate

name for his force, the "National State Defense Force Association," attracted the attentions of "B.J." (most likely, Maj. Gen. Bruce Jacobs) at the NGA. Holloway's group had no connections to the State Defense Force Association and was the only battalion in the only regiment of the Texas Reserve Militia.

50. "Texas State Guard May Have to Fight for Fiscal Survival," *Houston Post*, Apr. 28, 1985, sec. D, p. 7.
51. Mike Carter, "State Guard Reorganizes as 'Wackos' Infiltrate," *Salt Lake City Tribune*, Nov. 22, 1987, sec. B, pp. 1–2.
52. Ibid.
53. Carter, "State Guard's Aggression Forced a Change," *Salt Lake City Tribune*, Nov. 23, 1987, sec. B, p. 1.
54. Carter, "State Guard Reorganizes," sec. B, p. 2.
55. Carter, "State Guard's Aggression Forced a Change," sec. B, p. 1.
56. Maj. Gen. John L. Mathews to Maj. Gen Bruce Jacobs, Oct. 22, 1990, at the NGA Library, Washington, D.C.
57. Ed Connolly, "Militia Gets Riot Role: State Quietly Prepares Guard Backup," *San Jose Mercury News*, sec. A, May 2, 1989, pp. 1, 9.
58. George J. Stein, "Introductory Comments on Employment of State Defense Forces," notes from speech given at the SDFAUS conference in Kansas City, Mo., on Oct. 29, 1988. Notes at the NGA Library, Washington, D.C.
59. Ibid.
60. The movie *Red Dawn* is mentioned again and again by people who were drawn to the SDF movement in the 1980s to prepare for guerilla fighting against Communists in World War III.
61. Letter, Larry L. Larimer to "Mail Call," *American Survival Guide: The Magazine for Safer Living*, May, 1987, p. 18.
62. Margaret Miller, "Yankee Doodle Dandies . . . or Doodle Duds: The California Militia Awaits the Call," *California Journal* 23 (Aug., 1992): 382. Miller's article, written after most of the less controllable elements had been expelled, portrayed the force as ridiculous rather than dangerous.
63. Connolly, "Militia Gets Riot Role."
64. Maj. Gen. Joseph W. Griffin, *Annual Report 1990* (n.p.: Georgia Department of Defense, [1991]), p. 24.
65. Maj. Gen. James F. Fretterd, *Maryland National Guard Annual Report Fiscal Year 1991* (n.p., [1992]), p. 31.
66. Allen A. Moff, "State Defense Forces: A Volunteer Militia That Kicks In When the National Guard Is Federalized," *American Survival Guide*, Apr., 1991, pp. 36–38.
67. An example is Ed Connolly's "A Highly Irregular Force: Scandals of the State Militias," *Nation*, Mar. 18, 1991, pp. 338–42.
68. Ibid.
69. Jack Anderson, "State Militias Are Without Function," *Santa Cruz Sentinel*, Nov. 21, 1991, sec. A, p. 11.

70. Ibid.
71. T. C. Brown, "Militia Raising Concerns," *Cleveland Plain Dealer,* May 26, 1991, sec B, pp. 1, 4.
72. "State Civilian Militias Come to Attention: Some Say Gulf War Shows Need For Backup," *Cleveland Plain Dealer,* May 27, 1991, sec. C, pp. 1, 3.
73. Marilyn Sadler, "The Volunteer State Militia," *Memphis Flyer,* Aug. 1–14, 1993, p. 10.
74. Jerry Fogel stepped down from the leadership of the association but remained a member. He later became an officer in the New York Guard. From Fogel interview of Oct. 4, 2000.
75. Colonel McHenry had served in the Maryland Army National Guard as an artillery officer. He left the National Guard in 1963 after fifteen years of service because of the demands of his civilian career. He was recruited into the Maryland SDF by a friend in 1984.
76. Paul T. McHenry, Jr., "What's in the Name?" *Militia Journal* 1 (Dec., 1992): 6.
77. "Association By-Laws Changes," *Militia Journal* 2 (Winter, 1993): 5.
78. Maj. Carl E. Warner, Public Information Officer, TNDF, "Tornado Hits Lenoir City—TNDF Responds," *Militia Journal* 2 (Fall, 1993): 6.
79. Brig. Gen. Rafael L. Javis, "The Use of the State Guard in a Supporting Role to the National Guard—A Look at What Can Be Accomplished," *Militia Journal* 3 (Fall, 1994): 7–8.
80. "1993 Annual Meeting in Maryland," *Militia Journal* 2 (Winter, 1993): 3.
81. "FEMA Deputy Director Focuses on `Partnership,'" *Militia Journal* 4 (Spring, 1995): 1.
82. In summer 1994, the SGAUS listed the following eighteen states as having local chapters affiliated with the association: Alabama, Alaska, California, Maryland, Massachusetts, Michigan, Mississippi, New Mexico, North Carolina, New York, Ohio, Oklahoma, Oregon, Puerto Rico, South Carolina, Tennessee, Texas, and Virginia. In addition, chapters were in the process of being formed in the following fourteen states: Arizona, Colorado, Florida, Georgia, Illinois, Indiana, Louisiana, Nevada, Pennsylvania, Rhode Island, Utah, Vermont, Washington, and West Virginia.
83. Telephone interview with Col. Paul T. McHenry, Executive Director SGAUS, Jan. 21, 1999.
84. "Semantic Infiltration: To Be Militia or Not to Be," *SGAUS Journal* 5 (Summer, 1996): 6.
85. Ibid., p. 7.
86. Dan Walters, "California Has Its Own Militia," *Sacramento Bee,* May 2, 1995, sec. A, p. 3. Walters exposed an antigovernment steak of his own. His column ended with his observation that "[t]he militia should have been abolished years ago but like so many things the government does, it has continued to exist decades after its original purpose, if there ever was one, expired." Sentiments like this were common in the California press and not unheard of in other states.

87. "Survivalist Directory," *American Survival Guide: The Magazine of Self Reliance* 11 (July, 1989): 69. Susan Ferriss, "Assembly Will Rule On Greater Power For Militia," *San Francisco Examiner* (1995), in folder labeled "Militia—State Organizations," at the NGA Library, Washington, D.C.
88. "Official U.S. Delegation to Poland for V-E Day Headed by South Carolina State Guard," *Journal* 4 (Summer/Fall, 1995): 2.
89. Ibid., p. 5.
90. Col. Paul T. McHenry, "From the Desk of the Executive Director," *SGAUS Journal* 5 (Spring, 1996): 5, and Maj. Gen. Andre Trudeau, *The Rhode Island National Guard Fiscal Year 1994* (n.p., [1995]), p. 36. General Trudeau, the adjutant general for Rhode Island, mentions the existence of six chartered and six non-chartered companies. Unlike his predecessor in the 1920s, General Trudeau supported the companies and respected the long lineages they held. The companies participated in ceremonies throughout New England.
91. Tom Stuckey, "Maryland Militia a Bargain Force," *Washington Times*, insert "Metropolitan Times," Nov. 19, 1996, sec. C, p. 6. Although Stuckey's article was sympathetic to the Maryland Defense Force, the author still believed he needed to begin his story by explaining that the "Maryland militia" was not "storing caches of high-powered weapons or huddled in remote cabins making bombs."
92. Maj. Gen. James F. Fretterd, "Address to the State Guard Association," *Militia Journal* 3 (Spring, 1994): 1.
93. "Maryland: Good News—Bad News," *SGAUS Journal* 5 (Winter, 1996): 18.
94. "Notice of Annual Meeting Proposed By-Law Amendments and Resolutions," Sept. 16, 1996. Mailed to SGAUS members.
95. "NBC Dateline Covers the Meeting," *SGAUS Journal* 5 (Winter, 1996): 7.
96. NBC News, "Friendly Fire," *Dateline NBC*, June 17, 1997. Transcript provided by Burrelle's Information Services.
97. Ibid.
98. Telephone interview with Colonel McHenry, Jan. 21, 1999.
99. Telephone interview with Col. Paul T. McHenry, Nov. 27, 2000.
100. Few patterns are discernable between states that created an SDF in the 1980s and states that did not. In general, states with a strong militia tradition, such as Massachusetts, Texas, New York, and the Commonwealth of Puerto Rico, maintained one, whereas Montana did not. Creation of a State Guard during World War II was not much of a factor, as in the case of Nevada, which had no State Guard during the war but created one in the 1980s, while Hawaii and North Dakota did not.

Conclusion

1. Patrick Todd Mullins, "The Militia Clause, the National Guard, and Federalism: A Constitutional Tug of War," *George Washington Law Review* 57 (Dec., 1988): 328–29.

Bibliography

Archives

National Archives I, Washington, D.C.:

Records Group 165, War Department records pertaining to the State Guard, 1941–1942. Boxes 470–73.

National Archives II, College Park, Md.:

Records Group 160, War Department records pertaining to the State Guard, 1942–1948. Boxes 9–10.

Records Group 168, War Department records pertaining to the State Guard of the states. Boxes 104, 106–8, 115–21, 123–27, 134, 157–59.

Alabama State Archives, Birmingham:

SG 12261, Selective Service System: Home Guard.

RC2: G107, Home Defense Guards.

RC2: G122, National Guard.

Carlisle Barracks, Pennsylvania:

Hugh A. Drum Papers. Box 27, covering the years 1943–48, when General Drum served as Adjutant General of New York.

Georgia State Archives, Atlanta:

Records Group 22-2-10, Miscellaneous files of the governor.

Records Group 22-3-54, Miscellaneous files of the Georgia State Guard, 1940–1949.

Mississippi State Archives, Jackson:

Record Group 33, Adjutants General's Records.

Mississippi State Area Command (headquarters of the Mississippi National Guard), Jackson:

Drawer "General files 300.6-325.4 1898–1939."

File cabinet labeled "Miss Mil Dept 278 . . . ," with a fourth unreadable digit. (Also has "Col. Pleasant" taped on the front.)

Concord (Mass.) Free Public Library:

Collection of Materials Relating to the Concord Company of the Massachusetts State

Guard 1917. "A Sketch of the Concord Company, Massachusetts State Guard. 1917–1921." Typed manuscript, 1922.
"Scrap Book Events in Concord, Vol. VI 1937–1944."

National Guard Association Library, Washington, D.C.:
Two unlabeled boxes of files on State Defense Forces.

Government Publications

"Adjutant General's Office [Puerto Rico]. Annual report. . . ." 1947.
Ahner, Alfred F. "Report of the Adjutant General [of Indiana], Fiscal Year 1972." [1973].
———. *Annual Report Fiscal '83*. The Adjutant General's Office Indiana Army and Air National Guard, [1984].
Alabama Military Department, Office of the Adjutant General. *Annual Report, 1 Oct 87–30 Sep 88: The Alabama Reunion "Surprise Us" 1989*. N.p., [1989].
———. *Annual Report, 1 Oct 88–30 Sep 89*. N.p., [1989].
———. *Quadrennial Report of the Adjutant General for the Four Year Period Ending September 30, 1942*. Montgomery: State of Alabama, 1942.
Alabama State Military Department, Office of the Adjutant General. *Annual Report of the Adjutant General for Fiscal Year 1944–1945*. Montgomery: State of Alabama, 1945.
Alabama State Military Department, Office of the Adjutant General. *Annual Report of the Adjutant General for Fiscal Year Ending 30 September 1948*. Montgomery: State of Alabama, 1948.
Annual Report of the Adjutant General of the Commonwealth of Massachusetts for the Year Ending December 31, 1917. Boston: Wright & Potter Printing Co., State Printers, 1918.
Annual Report of the Adjutant General of the Commonwealth of Massachusetts for the Year Ending December 31, 1918. Boston: Wright & Potter Printing Co., State Printers, 1919.
Annual Report of the Adjutant General of the Commonwealth of Massachusetts for the Year Ending December 31, 1919. Boston: Wright & Potter Printing Co., State Printers, 1920.
Annual Report of the Adjutant General of the Commonwealth of Massachusetts for the Year Ending December 31, 1920. Boston: Wright & Potter Printing Co., State Printers, [1921].
Annual Report of the Adjutant General Puerto Rico National Guard Fiscal Year 1987–88. N.p., n.d.
Annual Report of the Chief of the Militia Bureau. Washington, D.C.: Government Printing Office, 1917.
Annual Report of the Chief of the Militia Bureau. Washington, D.C.: Government Printing Office, 1918.
Annual Report of the Chief of the Militia Bureau. Washington, D.C.: Government Printing Office, 1919.

Annual Report of the Chief of the Militia Bureau. Washington, D.C.: Government Printing Office, 1920.

Annual Report of the Chief of the Militia Bureau. Washington, D.C.: Government Printing Office, 1921.

Arndt, Theodore A. *Biennial Report of the Adjutant General of South Dakota July 1, 1944 to June 30, 1946.* N.p., n.d.

———. *Biennial Report of the Adjutant General South Dakota 1 July 1946 to 30 June 1948.* N.p., n.d.

Beckwith, Edward A. *Biennial Report of the Adjutant General of South Dakota July 1, 1940 to June 30, 1942.* N.p., [1942].

———. *Biennial Report of the Adjutant General of South Dakota July 1, 1942 to June 30, 1944.* N.p., [1944].

Berry, Charles W. *Annual Report of the Adjutant General for the Year 1919.* Albany: J. B. Lyons Company, 1921.

Biennial Report of the Adjutant General of the State of Delaware for the Two years Ending December 31, 1918. Milford, Del.: Milford Chronicle Publishing Co., 1919.

Biennial Report of the Adjutant General of the State of Delaware for the Two years Ending December 31, 1920. Milford, Del.: Milford Chronicle Publishing Co., 1919.

Biennial Report of the Adjutant General State of Tennessee Nashville, March 19, 1917. Nashville, [1917].

Biennial Report of the Adjutant General of the State of Tennessee for the Period January 1, 1917 to January 1, 1919. N.p., [1919].

Biennial Report of the Adjutant, Inspector, and Quartermaster General of the State of Vermont for the Two years Ending June 30, 1942. Burlington: Free Press Printing Co., [1942].

Biennial Report of the Adjutant, Inspector, and Quartermaster General of the State of Vermont for the Two years Ending June 30, 1944. Burlington: Free Press Printing Co., [1944].

Biennial Report of the Adjutant, Inspector, and Quartermaster General of the State of Vermont for the Two years Ending June 30, 1946. Burlington: Free Press Printing Co., [1946].

Biennial Report of the Adjutant, Inspector, and Quartermaster General of the State of Vermont for the Two years Ending June 30, 1948. Burlington: Free Press Printing Co., [1948].

Biennial Report Fifty-Seventh Biennial Report Period 1 July 1966–30 June 1968. The Nebraska Military Department, 1968.

Biennial Report of the Military Department State of Washington Office of the Adjutant General. Olympia: State Printing Plant, 1943.

Biennial Report of the Military Department State of Washington Office of the Adjutant General. Olympia: State Printing Plant, 1945.

Brown, Ames T. *Annual Report of the Adjutant General [of New York] For the Year 1941.* N.p., 1942.

———. *Annual Report of the Adjutant General [of New York] For the Year 1942.* Albany: Williams Press, 1943.

———. *Annual Report of the Adjutant General [of New York] For the Year 1943.* Albany: Williams Press, 1944.
———. *Annual Report of the Adjutant General [of New York] For the Year 1944.* Albany: Williams Press, 1945.
———. *Annual Report of the Adjutant General [of New York] For the Year 1945.* Albany: Williams Press, 1946.
———. *Annual Report of the Adjutant General [of New York] For the Year 1947.* New York: Publishers Printing Co, 1948.
Clark, Harvey C. *Report of the Adjutant General of Missouri January 1, 1917–December 31, 1920.* Jefferson City, 1920.
Cleere, George L. "Report of the Adjutant General of Alabama for the Fiscal Year 1942–1943." Montgomery: State of Alabama, 1942.
Collins, Vivian. *Report of the Adjutant General of the State of Florida For the Years 1939 and 1940.* N.p., [1941].
———. *Report of the Adjutant General of the State of Florida For the Years 1941 and 1942.* N.p., [1943].
———. *Report of the Adjutant General of the State of Florida For the Years 1943 and 1944.* N.p., [1945].
Couris, John J. *History Massachusetts State Guard.* N.p. [1976?].
Equal to the Task: A History of the Texas State Guard. A pamphlet prepared by the Office of the Adjutant General of the State of Texas. N.p., n.d.
Esteves, Luis Raul. *Annual Report of the Puerto Rico State Guard (Confidential) for the Period Ending June 30, 1942.* San Juan: Bureau of Supplies, Printing, and Transportation, 1942.
———. *Annual Report of the Puerto Rico State Guard for the Period Ending June 30, 1943.* San Juan: Insular Procurement Office Printing Division, 1943.
Everett, Tim, ed. *Report of the Adjutant General of Michigan, 1983–84.* Lansing: Allied Printing, 1984.
Farmer, Arthur J. *Mississippi Military Department Annual Report 1 July 1987–30 June 1988.* Jackson, 1988.
Flemming, Raymond H. *Report of the Adjutant General of the State of Louisiana January 1, 1940, to December 31, 1941 to the Governor and Legislature of the State of Louisiana.* N.p., [1942].
———. *Biennial Reports, 1944–1945, Department of Military Affairs and Selective Service System State of Louisiana.* N.p., [1946].
———. *Biennial Reports, 1946–1947, Department of Military Affairs and Selective Service System State of Louisiana.* N.p., [1948].
Frazier, T. A. *From Peace to War: State of Tennessee Consolidated Report of the Adjutant General for the Period January 16, 1939–January 4, 1943.* N.p., [1943].
Fretterd, James F. *Maryland National Guard Annual Report Fiscal Year 1990.* N.p., [1991].
———. *Maryland National Guard Annual Report Fiscal Year 1991.* N.p., [1992].
Garner, James H. *Annual Report of the Adjutant General for 1 July 91–30 June 92.* Jackson, Miss., 1992.

Grayson, Thomas J. *Biennial Report of the Adjutant General State of Mississippi from July 1, 1939–June 30, 1941.* Jackson, 1941.

Griffin, Joseph, W. *Annual Report 1985.* Georgia Department of Defense, [1986].

———. *Annual Report for Fiscal Year 1990.* Georgia Department of Defense, [1991].

Hancock, James E. "Report of the Adjutant General State of Montana." [1942].

Haskett, George M. *Biennial Report of the Adjutant General of Washington July 1, 1960 to June 30, 1962.* State of Washington Military Department, 1962.

Hausauer, Karl F. *Annual Report of the Chief of Staff to the Governor For the Division of Military and Naval Affairs for the Year 1950.* MG Karl F. Hausauer, Chief of Staff to the Governor. Albany: Williams Press, 1951.

———. *Annual Report of the Chief of Staff to the Governor For the Division of Military and Naval Affairs for the Year 1951.* Albany: Williams Press, 1951.

———. *Annual Report of the Chief of Staff to the Governor For the Division of Military and Naval Affairs for the Year 1954.* New York: Herald Square Press, 1955.

———. *Annual Report of the Chief of Staff to the Governor For the Division of Military and Naval Affairs for the Year 1955.* New York: Herald Square Press, 1955.

———. *Annual Report of the Chief of Staff to the Governor For the Division of Military and Naval Affairs for the Year 1956.* N.p., n.d.

Hays, Ralph. *Biennial Report of the Adjutant General of the State of Mississippi from July 1, 1941–June 30, 1943.* Jackson, 1943.

———. *Biennial Report of the Adjutant General of the State of Mississippi from July 1, 1943–June 30, 1945.* Jackson, 1945.

Heywood, Edwin W. *Annual Report.* Department of the Adjutant General State of Maine, 1970.

Historical Background of the New York Guard. A history prepared by the Office of the Adjutant General of the State of New York. N.p., n.d.

History—Maine State Guard. A report prepared by the Office of the Adjutant General of the State of Maine, 1946.

Hodges, Warren D. *Annual Report Fiscal Year 1983 Maryland National Guard.* N.p., [1984].

Jones, Earle M. *Biennial Report of the Adjutant General For the Period 1 July, 1950 to 30 June 1952.* N.p., [1952].

Kearney, George, ed. *Official Opinions of the Attorneys General of the United States Advising the President and Heads of Departments in relation to Their Official Duties.* Washington, D.C.: Government Printing Office, 1913.

Lance, Mark W. *Report of the Adjutant General of the State of Florida For the Years 1947 and 1948.* N.p., [1949].

———. *Report of the Adjutant General of the State of Florida for the Years 1951 and 1952.* N.p., [1952].

McLane, Curren R. *A History of the Texas State Guard.* Headquarters, Texas State Guard. N.p., n.d.

McLean, M. R. "Military Laws of Kansas Pertaining to the Kansas State Guard AG's office, Topeka, 1 February, 1946."

Makinney, F. W. *Annual Report of The Adjutant General Territory of Hawaii July 1, 1946, to June 30, 1947.* The Printshop of Hawaii Co., 1942.
Marchant, T. Eston. *Report of the Adjutant General of South Carolina fiscal Year 1 July 1981–30 June 1982.* Printed Under the Direction of the State Budget and Control Board, n.d.
———. *Report of the Adjutant General of South Carolina 1 July 1983–30 June 1984.* N.p., n.d.
———. *South Carolina Office of the Adjutant General Annual Report 1993–1994.* Printed under the direction of the State Budget and Control Board, n.d.
Maryland, Council of Defense, *Report of the Maryland Council of Defense to the Governor and the General Assembly, 1919–1920.* N.p., 1920.
Melhouse, La Clair A. *Biennial Report [North Dakota] 1 July 1964–30 June 1966.* N.p., 1966.
Militia Act of 1792: An Act more effectually to provide for the National Defence by establishing a Uniform Militia throughout the United States. The Statutes at Large of the United States of America. Vol. 1, pp. 271–74. Boston: Charles C. Little and James Brown, 1848.
Militia Act of 1903: An Act To promote the efficiency of the militia, and for other purposes. Statutes at Large of the United States of America. Vol. 32, part 1, pp. 774–80. Washington, D.C.: U.S. Government Printing Office, 1904.
Militia Act of 1908: An Act To further amend the Act entitled "An Act To promote the efficiency of the militia, and for other purposes, approved January twenty-first, nineteen hundred and three". Statutes at Large of the United States of America. Vol. 35, part 1, pp. 399–403 Washington, D.C.: U.S. Government Printing Office, 1909.
Moran, George C. *Biennial Report of the Adjutant General of Michigan For the Period 1 July 1948 Through 30 June 1950 (with Addenda to 15 May 1951).* Multilith Section, Adjutant General's Office, 1951.
———. *Report of the Adjutant General 1952.* N.p., 1953.
Morrill Act: An Act donating Public Lands to the several States and Territories which may provide Colleges for the benefit of Agriculture and the Mechanic Arts. Statutes at Large of the United States of America. Vol. 13, Boston: Charles C. Little and James Brown, 1863.
National Defense Act Approved. Washington, D.C.: Government Printing Office, 1927.
National Defense Act of 1916: An Act for making further and more effectual provisions for the National Defense, and for other purposes. Statutes at Large of the United States of America. Vol. 39, part 1, pp. 166–217. Washington, D.C.: U.S. Government Printing Office, 1917.
1974 Annual Report State of New York Division of Military and Naval Affairs. N.p., n.d.
O'Keefe, John. *Biennial Report of the Adjutant General of the State of Mississippi from July 1, 1937 to June 30, 1939.* Jackson, 1939.
O'Sullivan, C. D. *Biennial Report of the Adjutant General July 1, 1946 to June 30, 1948.* N.p., [1948].
———. *Biennial Report of the Adjutant General July 1, 1948 to June 30, 1950.* N.p., [1950].

Pancoast, D. F. *Report of the Adjutant General of Ohio to the Governor of the State of Ohio for the Year 1943*. Columbus: Heer Printing Co, 1944.

Pearson, Le Roy. *Report of the Adjutant General of Michigan for the Period January 1, 1943, to June 30, 1944*. N.p., [1944].

"Public Document Number 49," *Fourteenth Annual Report of the Police Commission for the City of Boston: Year Ending November 30, 1919*. Boston: Wright and Potter Printing Co., State Printers, 1920.

Puerto Rico National Guard. *Annual Report of the Adjutant General Fiscal Year 1984–85*. Puerto Rico National Guard, n.d.

———. *Annual Report of the Adjutant General Fiscal Year 1985–86*. Puerto Rico National Guard, n.d.

Regulations For the Connecticut Home Guard. Published by the (Conn.) Military Emergency Board, 1918.

Report of the Adjutant General of the State of Connecticut to the Governor and the Assembly for the Period Jan. 1, 1940 to Dec 31, 1940. N.p., [1941].

Report of the Adjutant General of the State of Connecticut to the Governor and the General Assembly for the Period Jan. 1, 1941 to Dec. 31, 1941. N.p., [1942].

Report of the Adjutant General [of the State of Connecticut] to the Governor for the Period January 1, 1942 to December 31, 1944. N.p., [1945].

Report of the Adjutant General State of Maryland for 1918–1919. Annapolis: Weekly Advertizer, [1920].

Reports of the Adjutant General of New Jersey, 1916–1918. Union Hill: Hudson Printing Company, 1919.

Reports of the Adjutant General of New Jersey, 1918–1920. Trenton: Published by the State, 1920.

Reports of the Adjutant General of New Jersey, 1921. Rahway: Reformatory Print, 1921.

Report of the Adjutant General of the Commonwealth of Virginia for the Year Ending December 31, 1917. Richmond: Davis Bottom, Superintendent of Public Printing, 1918.

Report of the Adjutant General of the Commonwealth of Virginia for the Year Ending December 31, 1918. Richmond: Davis Bottom, Superintendent of Public Printing, 1919.

Report of the Adjutant General of the Commonwealth of Virginia for the Year Ending December 31, 1919. Richmond: Davis Bottom, Superintendent of Public Printing, 1920.

Report of the Adjutant General of the Commonwealth of Virginia for the Year Ending December 31, 1920. Richmond: Davis Bottom, Superintendent of Public Printing, 1921.

Report of the Adjutant General of Washington, 1917–1918. Olympia: Frank M. Lamborn Public Printer, 1918.

Report of the Adjutant General of West Virginia, 1940–1942. Charleston, W.V.: Jarrett Printing Company, [1943].

Report of the Adjutant General of West Virginia, 1943–1944. Charleston, W.V.: Jarrett Printing Company, [1945].

Report of the Adjutant General of West Virginia, 1945–1946. Charleston, W.V.: Jarrett Printing Company, [1947].
Report of the Chief of the National Guard Bureau Fiscal Year Ending 30 June 1939. Washington, D.C.: Government Printing Office, 1946.
Report of the Chief of the National Guard Bureau Fiscal Year Ending 30 June 1940. Washington, D.C.: Government Printing Office, 1946.
Report of the Chief of the National Guard Bureau Fiscal Year Ending 30 June 1941. Washington, D.C.: Government Printing Office, 1946.
Report of the Chief of the National Guard Bureau Fiscal Year Ending 30 June 1946. Washington, D.C.: Government Printing Office, 1946.
Report of the Minnesota Commission of Public Safety, 1917–18. A report prepared by the Minnesota Department of Labor and Industry. St. Paul: Louis F. Dow Company, 1919.
Scales, Erie C. *Biennial Report of the Adjutant General State of Mississippi from January 1, 1916–December 31, 1917.* Jackson: Tucker Printing, 1917.
———. *Biennial Report of the Adjutant General of the State of Mississippi for the Years 1918–1919 to the Governor.* Jackson: Tucker Printing House, 1920.
Report of South Dakota State Council of Defense May 8, 1917 to Dec. 31, 1919. N.p., n.d.
"Report to the Secretary of War on the Activities of the State Guards for the Year Ending October 31, 1941." National Guard Association Library. Washington, D.C.
Seventeenth Biennial Report of the Adjutant General of the State of Oregon To the Governor and Commander-in-Chief For the Period November 1, 1918, to October 31, 1920. Salem, Ore.: State Printing Department, 1921.
Sixteenth Biennial Report of the Adjutant General of the State of Oregon To the Governor and Commander-in-Chief For the Period November 1, 1916, to October 31, 1918. Salem, Ore.: State Printing Department, 1919.
Smoot, P. M. *Annual Report of The Adjutant General Territory of Hawaii July 1, 1940, to June 30, 1941.* Hawaii Printing Co., 1942.
———. *Annual Report of The Adjutant General Territory of Hawaii July 1, 1941, to June 30, 1942.* Hawaii Printing Co., 1942.
State Defense Force Manual. Harrisburg, Pa.: Military Service Publishing Co., 1941.
State of Connecticut Report of the Adjutant General Two Years Ended September 30, 1918. Hartford: Published by the State, 1919.
State of Louisiana. *Military Code Providing for the Organization, Discipline, and Maintenance of the National Guard and Other Military Forces of the State Being Act 164 of 1940 As Amended and Related Laws.* New Orleans: Department of Military Affairs, 1946.
State of Rhode Island and Providence Plantations. *Annual Report of the Adjutant General and Quartermaster General of the State of Rhode Island for the Year 1917.* Providence: Oxford Press, n.d.
———. *Annual Report of the Adjutant General and Quartermaster General of the State of Rhode Island for the Year 1918.* Providence: E. L. Freeman Company, n.d.
———. *Annual Report of the Adjutant General of the State of Rhode Island for the Year 1941.* N.p., n.d.

———. *Annual Report of the Adjutant General of the State of Rhode Island for the Year 1943*. N.p., n.d.
———. *Annual Report of the Adjutant General of the State of Rhode Island for the Year 1944*. N.p., n.d.
———. *Annual Report of the Adjutant General of the State of Rhode Island for the Year 1945*. N.p., n.d.
———. *Annual Report of the Adjutant General of the State of Rhode Island for the Year 1947*. N.p., n.d.
State of Washington Military Department. *Biennial Report of the Adjutant General July 1, 1972–June 30, 1974*. N.p., n.d.
Sullivan, Maurice J. *State of Nevada Biennial Report of the Adjutant General 1917–1918*. Carson City: State Printing Office, 1919.
Thirtieth Biennial Report of The Military Department of the State of Oregon to the Governor For the Period November 1, 1945, to December 31, 1946. N.p., n.d.
Thirty-first Biennial Report of The Military Department of the State of Oregon to the Governor For the Period January 1, 1947, to December 31, 1948. State Printing Department, n.d.
Thirty-second Biennial Report of the Military Department of the State of Oregon to the Governor for the Period January 1, 1949 to December 31, 1950. State Printing Department, n.d.
Thirty-third Biennial Report of the Military Department of the State of Oregon to the Governor for the Period January 1, 1951 to December 31, 1952. State Printing Department, n.d.
To Amend the National Defense Act of 1916. Washington, D.C.: Government Printing Office, 1927.
Twenty-eighth Biennial Report of The Military Department of the State of Oregon to the Governor For the Period November 1, 1940, to October 31, 1942. State Printing Office, n.d.
Twenty-ninth Biennial Report of The Military Department of the State of Oregon to the Governor For the Period November 1, 1942, to October 31, 1944. State Printing Department, n.d.
Twenty-first Biennial Report of the Adjutant General of the State of Kansas, 1917–1918. Topeka: Kansas State Printing Plant, 1919.
Twenty-second Biennial Report of the Adjutant General, 1997–1920. Topeka: Kansas State Printing Plant, 1921.
Twenty-seventh Biennial Report of The Military Department of the State of Oregon to the Governor For the Period November 1, 1938, to October 31, 1940. State Printing Office, n.d.
United States Army. Western Defense Command. "Western Defense Command Counter Fifth Column Plan, 1944" Mimeo, July 14, 1944.
United States Department of the Army, *Regulations for State Guard*. AR 915-10 Washington, D.C.: War Department, May 14, 1951.
United States National Guard Bureau. *R-1480 War Department Policies Regarding State Guard*. Washington D.C., 1947. Directives & Reports.

United States Senate Committee on Military Affairs. *To Amend the National Defense Act, to Create the Reserve Division of the War Department.* Washington, D.C.: Government Printing Office, 1935.

———. *To Amend the National Defense Act of June 3, 1916, as Amended: Report (to accompany S 4026).* Washington, D.C.: Government Printing Office, 1936.

———. *Amending the National Defense Act of June 3, 1916, as Amended by Act of June 6, 1924.* Washington, D.C.: Government Printing Office, 1940.

———. *The Home Guard: Hearings before the Committee on Military Affairs United States Senate.* S 4175. Washington, D.C.: Government Printing Office, 1940.

———. *The Home Guard: Report to Accompany House Report 10495.* Washington, D.C.: Government Printing Office, 1940.

———. *Amending Sec. 61 of the National Defense Act June 1916: Report.* Washington, D.C.: Government Printing Office, 1941.

United States War Department. "Document Number 32: The Organized Militia of the United States," Washington, D.C.: Government Printing Office, 1897.

———. *The Progress of the War Department in Compliance with the National Defense Act of 1916.* Washington, D.C.: Government Printing Office, 1922.

———. *Bulletin 29.* Washington, D.C.: War Department, Sept. 3, 1941. "II Act of Congress—National Defense Act amended—Permission to organize military units not a part of the National Guard.—The following act of Congress (Public Law 214—77th Cong.) is published for the information and guidance of all concerned."

———. *Regulations for State Guard.* AR 850-250 Washington, D.C.: War Department, Apr. 21, 1941.

———. *Regulations for State Guard.* AR 850-250 Washington, D.C.: War Department, Aug. 9, 1943.

———. *Regulations for State Guard.* AR 850-250 Washington, D.C.: War Department, Dec. 13, 1945.

———. Service of Supply. *State Guard Property Accounting.* Washington, D.C.: Office of the Chief of the National Guard Bureau, Mar. 1, 1943.

———. "Recommended Training Program for State Guards, Medical Detachment." Mimeo, May 27, 1942.

———. "Directives and Reports on State Guards, 1942–1946." Policy file of 8 mimeo items. Circa 1946.

"Virgin Islands National Guard Annual Report Fiscal Year 1980, 1 October 1979–30 September 1980." [1980].

Volunteer Act of 1806: An Act authorizing a detachment from the Militia of the United States. Statutes at Large of the United States of America. Vol. 2, pp. 383–84. Boston: Charles C. Little and James Brown, 1845.

Volunteer Act of 1861: An Act To authorize the employment of volunteers to aid in enforcing the laws and protecting private property. Statutes at Large of the United States of America. Vol. 12, pp. 268–69. Boston: Charles C. Little and James Brown, 1863.

Volunteer Act of 1914: An Act To provide for raising the volunteer forces of the United States in time of actual or threatened war. Statutes at Large of the United States of America. Vol. 38, part 1, p. 347. Washington, D.C.: U.S. Government Printing Office, 1915.

Waller, S. Gardner. *The Report of the Adjutant General of the State of Virginia Division of Military Affairs For the Period January 1, 1940, to December 31, 1940.* Richmond: Division of Purchase and Printing, 1941.

———. *The Report of the Adjutant General of the State of Virginia Division of Military Affairs For the Period January 1, 1941, to December 31, 1941.* Richmond: Division of Purchase and Printing, 1942.

———. *The Report of the Adjutant General of the State of Virginia Division of Military Affairs For the Period January 1, 1942, to December 31, 1942.* Richmond: Division of Purchase and Printing, 1943.

———. *The Report of the Adjutant General of the State of Virginia Division of Military Affairs For the Period January 1, 1943, to December 31, 1943.* Richmond: Division of Purchase and Printing, 1944.

———. *The Report of the Adjutant General of the State of Virginia Division of Military Affairs For the Period January 1, 1944, to December 31, 1944.* Richmond: Division of Purchase and Printing, 1945.

———. *The Report of the Adjutant General of the State of Virginia Division of Military Affairs For the Period January 1, 1945, to December 31, 1945.* Richmond: Division of Purchase and Printing, 1948.

———. *The Report of the Adjutant General of the State of Virginia Division of Military Affairs For the Period January 1, 1946, to December 31, 1946.* Richmond: Commonwealth of Virginia Division of Purchase and Printing, 1948.

———. *The Report of the Adjutant General of the State of Virginia Division of Military Affairs For the Period January 1, 1947, to December 31, 1947.* Richmond: Commonwealth of Virginia Division of Purchase and Printing, 1949.

Warfield, Henry M. *Report of the Adjutant General State of Maryland for 1916–1917.* Baltimore: King Brothers State Printers, [1918].

White, Jay H. *State of Nevada Biennial Report of the Adjutant General for the Period July 1, 1940 to June 30 1942, Inclusive.* Carson City: State Printing Office, 1942.

———. *State of Nevada Biennial Report of the Adjutant General for the Period July 1, 1942 to June 30 1944, Inclusive.* Carson City: State Printing Office, 1944.

———. *Report of the Adjutant General of Nevada For the Fiscal Years Ending June 30 1945–1946.* Carson City: State Printing Office, 1946.

Wilson, William P. *Biennial Report of the Adjutant General of the State of Mississippi to the Governor and Legislature for the Biennium Ending June 30, 1949.* Jackson, 1949.

———. *Biennial Report of the Adjutant General of the State of Mississippi to the Governor and Legislature for the Biennium Ending June 30, 1951.* Jackson, 1951.

———. *Biennial Report of the Adjutant General of the State of Mississippi for the Biennium Ending 30 June 1953.* Jackson, 1953.

Books

Adams, Elliott B., and Dafferner, Frank T. *First Division Tennessee State Guard Training Manual 1943.* Nashville: T. A. Frazier, Major General, Commanding, 1943.

Allen, Gwenfread. *Hawaii's War Years, 1941–1945.* Honolulu: University of Hawaii Press, 1950.

Ambrose, Stephen E. *Upton and the Army.* Baton Rouge: Louisiana State University Press, 1964.

Bailey, Kenneth R. *Mountaineers Are Free: A History of the West Virginia National Guard.* St. Albens, W.V.: Harless Printing, 1978.

Beckwith, Edmund R. *Lawful Actions of State Military Forces.* New York: Random House, 1944.

Belfiglio, Valentine J. *Honor, Duty, Pride: A History of the Texas State Guard.* Austin: Eakin Press, 1995.

Binkin, Martin, and William W. Kaufman. *U.S. Army Guard and Reserve: Rhetoric, Realities, Risks.* Washington, D.C.: Brookings Institution, 1989.

Blum, John Morton. *V Was for Victory: Politics and American Culture During World War II.* New York: Harcourt Brace Jovanovich, 1976.

Breen, William J. *Uncle Sam at Home: Civilian Mobilization, Wartime Federalism, and the Council of National Defense, 1917–1919.* Westport, Conn.: Greenwood Press, 1984.

A Brief History of Troop A 107th Regiment of Cavalry Ohio National Guard The Black Horse Troop For Many Years Known as The First City Troop of Cleveland. Printed for the Active members and the Veterans' Association on the Occasion of the Opening of the New Armory, Cleveland, Ohio, 1923.

Brown, Roger Allen, William Fedorochko, Jr., and John F. Schank. *Assessing the State and Federal Missions of the National Guard.* Santa Monica, Calif.: Rand—National Defense Research Institute, 1995.

Clifford, John Garry. *The Citizen Soldiers: The Plattsburg Training Camp Movement, 1913–1920.* Lexington: University Press of Kentucky, 1972.

Coakley, Robert W. *The Role of Federal Military Forces in Domestic Disorders, 1789–1878.* Army Historical Series. Washington, D.C.: Center of Military History, 1988.

Coffman, Edward M. *The Old Army: A Portrait of the American Army in Peacetime, 1784–1898.* New York: Oxford University Press, 1986.

Coffman, Edward M. *The War to End All Wars: The American Military Experience in World War I.* Madison: University of Wisconsin Press, 1968.

Colby, Elbridge. *The National Guard of the United States: A Half Century of Progress.* Manhattan, Kans.: Military Affairs/Aerospace Historian Publishing, 1977.

Commonwealth of Pennsylvania. *Pennsylvania at War, 1941–1945.* Harrisburg: Pennsylvania History and Museum Commission, 1946.

———. *Pennsylvania's Second Year at War, December 7, 1942–December 7, 1943.* Harrisburg: Pennsylvania History and Museum Commission, 1946.

Cooper, Jerry M. *The Army and Civil Disorder: Federal Military Intervention in Labor Disputes, 1877–1900.* Westport, Conn: Greenwood Press, 1980.

———. *The Militia and National Guard in America since Colonial Times: A Research Guide.* Westport Conn.: Greenwood Press, 1993.

———. *The Rise of the National Guard: The Evolution of the American Militia, 1865–1920.* Lincoln: University of Nebraska Press, 1997.

Craig, George A. *History: I Massachusetts State Guard, II Massachusetts State Guard Veterans.* N.p. George A. Craig, 1931.

Crossland, Richard, B., and James T. Currie. *Twice the Citizen: A History of the United States Army Reserve, 1908–1983.* Washington, D.C.: Office of the Chief, Army Reserve, 1984.

Cress, Lawrence Delbert. *Citizens in Arms: The Army and Militia in American Society to the War of 1812.* Chapel Hill: University of North Carolina Press, 1982.

Cushing, John T., Arthur F. Stone, and Harold P. Sheldon, eds. *Vermont in the Great War, 1917–1919.* Burlington: Free Press Printing Company for the State of Vermont, 1928.

Cuneo, William H., and Leon de Valinger. *Delaware's Role in World War II, 1940–46.* Dover: N.p., 1955.

Daugherty, Robert. *Weathering the Peace: The Ohio National Guard in the Interwar Years, 1919–1940.* Lanham, Md.: Wright State University Press, 1992.

de Casenave, Luis A. *Historia De La Guardia Estatal De Puerto Rico (1508–1992): Resena Historica del Origen de La Guardia Estatal de Puerto Rico.* San Juan: Guardia Estatal de Puerto Rico, 1992.

Dees, Morris, and James Corcoran. *Gathering Storm: The Story of America's Militia Network.* New York: Harper Collins Publishing, 1996.

Derthick, Martha. *The National Guard in Politics.* Cambridge: Harvard University Press, 1965.

Dieges, Charles J., and T. F. Donovan, eds. *The Home Guard Manual: Embracing the Essential Parts of Citizen-Soldiery, Manual of Arms, and That Portion of the Field Regulation Relating to Military Police.* New York: Sherwood Company, 1917.

Dupuy, R. Ernest. *The National Guard: A Compact History.* New York: Hawthorn Books, 1971.

Eby, C. Arthur. *A Record of the Maryland State Guard.* Baltimore: N.p., 1920.

Editorial Staff of the War Records Committee Haverhill. *Haverhill in World War II.* Haverhill, Mass.: City of Haverhill, 1946.

Folwell, William Watts. *A History of Minnesota.* Vol. 3. Saint Paul: Minnesota Historical Society, 1926.

Fowles, Brian Dexter. *A Guardian in Peace and War: The History of the Kansas National Guard, 1854–1987.* Manhattan, Kans.: Sunflower University Press, 1989.

Fowles, Lloyd W. *An Honor to the State: The Bicentennial History of the First Company Governor's Foot Guard.* Hartford, Conn.: First Company Governor's Foot Guard, 1971.

Goodson, Felix E. *The Indiana Guard Reserve.* Indianapolis: Indiana Creative Arts, 1998.

Gibbs, Christopher C. *The Great Silent Majority: Missouri's Resistance to World War I.* Columbia: University of Missouri Press, 1988.

Guyol, Philip N. *Democracy Fights: A History of New Hampshire in World War II.* Hanover: Pennsylvania History and Museum Committee, 1945.
Hagedorn, Hermann. *Leonard Wood: A Biography.* New York: Harper & Brothers, 1931.
Hanson, Joseph Mills. *South Dakota in the World War, 1917–1919.* [Pierre?], S. Dak.: State Historical Society, 1940.
Hartman, Douglas R. *Nebraska's Militia: The History of the Army and Air National Guard, 1854–1991.* Nebraska Military Department. [Lincoln?], Nebr.: Donning Company/Publishers, 1994.
Hartzell, Karl D. *The Empire State at War: Wold War II.* New York: State of New York, 1949.
Hawk, Robert. *Florida's Army: Militia/State Troops/National Guard, 1565–1985.* Englewood, Fla.: Pineapple, 1986.
HERO. *U.S. Home Defense Forces Study.* Dunn Loring, Va.: Historical Evaluation & Research Organization, Prepared under contract for the Office of the Assistant Secretary of Defense, 1981.
Higginbotham, Don. *The War of American Independence: Military Attitudes, Policies, and Practices, 1763–1789.* Bloomington: Indiana University Press, 1977.
Higham, Robin, ed. *Bayonets in the Streets: The Use of Troops in Civil Disturbances.* Lawrence: University Press of Kansas, 1969.
Hill, Jim Dan. *The Minute Man in War and Peace: A History of the National Guard.* Harrisburg, Pa.: Stackpole Company, 1964.
History of the California State Guard. N.p., The Adjutant General of the State of California, 1946.
History of the Richardson Light Guard of Wakefield Mass. Wakefield: Item Press, 1926.
Holbrook, Franklin F., and Livia Appel. *Minnesota in the War with Germany.* 2 vols. St. Paul: Minnesota Historical Society, 1928 (vol. 1), 1932 (vol. 2).
Hollister, C. Warren. *Anglo-Saxon Military Institutions on the Eve of the Norman Conquest.* Oxford, England: Oxford University Press, 1962.
How Minnesota Gave to the United States The First Military Motor Corps. Minneapolis: Bancroft Printing Company, 1919.
Jacobs, James Ripley. *The Beginning of the U.S. Army, 1783–1812.* Princeton: Princeton University Press, 1947.
Johnson, Stephen D., and Gary S. Poppleton. *Cloth Insignia of the U.S. State Guards and State Defense Forces.* Hendersonville, Tenn.: Richard W. Smith, 1993
Kohn, Richard H. *Eagle and Sword: The Federalists and the Creation of the Military Establishment in America, 1783– 1802.* New York: Free Press, 1975.
Kohn, Richard H. *The U.S. Military under the Constitution.* New York: New York University Press, 1991.
Krenek, Harry. *The Power Vested: The Use of Martial Law and the National Guard in Texas, 1919–1932.* Austin: Presidial Press, 1980.
Kunz, Virginia Brainard. *Muskets to Missiles: A Military History of Minnesota.* Saint Paul: Minnesota Statehood Centennial Commission, 1958.

Lampe, William, ed. *Tulsa County in the World War: An Authorized History.* Tulsa: Tulsa County Historical Society, 1919.
Lane, Jack C. *Armed Progressive: General Leonard Wood.* San Rafael, Calif.: Presidio Press, 1978.
Larson, R. A. *Wyoming's War Years, 1941–1945.* Laramie: University of Wyoming, 1954.
Leblanc, Cyrille, and Thomas F. Flynn. *Gardner in World War II.* Gardner, Mass.: Hatton Press, 1947.
Lee, Alfred McClung, and Norman D. Humphrey. *Race Riot: Detroit, 1943.* New York: Octagon Books, 1968.
Link, Arthur S. *Wilson: Confusion and Crises, 1915–1916.* Princeton: Princeton University Press, 1964.
Listman, John W., Jr., Robert K. Wright, Jr., and Bruce D. Hardcastle, eds. *The Tradition Continues: A History of the Virginia National Guard, 1607–1985.* N.p., Taylor Publishing Company.
Mackenzie, S. P., *The Home Guard: A Military and Political History.* New York: Oxford University Press, 1995.
Maryland Historical Society, War Records Division. *Maryland in World War II: Volume 1, Military Participation.* Baltimore: Maryland Historical Society, 1950.
Mahon, John K. *History of the Militia and the National Guard.* New York: Macmillan, 1983.
McMillen, Neil R. *Dark Journey: Black Mississippians in the Age of Jim Crow.* Chicago: University of Illinois Press, 1989.
Miller, Duane Ernest. *Michigan's State Defense Forces: A Concise History and Lineage of the Michigan Emergency Volunteers and Its Predecessors.* Lansing, Mich.: Berlin Green Press, 1999.
National Rifle Association of America. *Practical Home Guard Organization for Reserve Militia or "Minute Men."* N.p.: National Rifle Association of America, [1942].
———. *Practical Home Guard Organization Supplementary Bulletin Covering Application of the "Minute Man" to Large Cities.* N.p.: National Rifle Association of America, [1942].
Putnam, Eben, ed., *Report of the Commission on Massachusetts' Part in the World War History* Vol. 1. Boston: Commonwealth of Massachusetts, 1931.
Rhodes, Gwen R., *South Carolina Army National Guard.* Dallas: Taylor Publishing Company, 1988.
Riker, William H. *Soldiers of the States: The Role of the National Guard in American Democracy.* Washington. D.C.: Public Affairs Press, 1957.
Russell, Francis. *A City in Terror: The 1919 Boston Police Strike.* New York: Viking Press, 1975.
Salisbury, C. A. *Soldiers of the Mists: Minutemen of the Alaska Frontier.* Missoula: Pictorial Publishing Company, 1992.
Schlegel, Marvin Wilson. *Virginia on Guard: Civilian Defense and the State Militia in the Second World War.* Richmond: Virginia State Library, 1949.

Shy, John W. *A People Numerous and Armed: Reflections on the Military Struggle for American Independence.* Ann Arbor: University of Michigan Press, 1990.
Sligh, Robert Bruce. *The National Guard and National Defense: The Mobilization of the Guard in World War II.* New York: Praeger, 1992.
South Dakota in World War II. Vol. 1. N.p.: World War II Hist Commission, n.d.
Spector, Ronald H. *Eagle Against the Sun: The American War with Japan.* New York: Vintage Books, 1985.
Stambaugh, John L. *Company 6-F Berks County Pennsylvania State Guard Reserve of Rehrersburg, PA.* Rehrersburg, Pa.: J. L. Stambaugh, 1959.
State of New York. *The New York Red Book 1942: An Illustrated Publication Containing Authentic Information Relating to the Executive, Legislative, Judicial and Political Affairs of the State.* Albany: Williams Press, 1942.
State of Washington Military Department Office of the Adjutant General. *A Brief History of the National Guard of Washington.* State of Washington, 1957.
Stone, Richard G., Jr. *A Brittle Sword: The Kentucky Militia, 1776–1912.* Lexington: University Press of Kentucky, 1977.
The Two Hundred and Seventy-Ninth Annual Record of the Ancient and Honorable Artillery Company of Massachusetts, 1916–1917. Boston: George E. Crosby Co., 1918.
Upton, Emory. *The Military Policy of the United States.* Washington: Government Printing Office, 1907.
Watt, William J., and James R. H. Spears, eds. *Indiana's Citizen Soldiers: The Militia and National Guard in Indiana History.* Indianapolis: Indiana State Armory Board, 1980.
Weigley, Russell F. *History of the United States Army.* New York: Macmillan, 1967.
Wright, Paul R. *Training the Organized Reserves under the National Defense Act.* Chicago: 1924.
Zack, Charles S. *Holyoke in the Great War.* Holyoke, Mass.: Transcript Publishing Company, 1919.

Articles, Dissertations, and Theses

Ansell, Samuel T. "Legal and Historical Aspects of the Militia." *Yale Law Review* 26 (Apr., 1917): 471–80.
Bailey, Kenneth Roy. "A Search for Identity: The West Virginia National Guard, 1877–1921." Ph.D. diss. Ohio State University, 1976.
Bittle, George Cassel. "In Defense of Florida: The Organized Florida Militia from 1821 to 1920." Ph.D. diss. Florida State University, 1965.
Brittain, William J. "Home-Front Guardians." *American Legion Magazine* 36 (Mar., 1944): 20, 29–30, 32, 34.
Beaumont, Roger A. "Constabulary or Fire Brigade? The Army National Guard." *Parameters* 80 (Sept., 1984): 62–69.
Blumenson, Martin. "On the Function of the Military in Civil Disorders." In Roger W. Little, ed., *Handbook of Military Institutions.* Beverly Hills, Calif.: Sage Publications, 1971, pp. 500–25.

Chamberlain, Robert S. "The Northern State Militia." *Civil War History* 4 (June, 1958): 105–18.

Christensen, Lawrence O. "World War I in Missouri Part 2" *Missouri Historical Review* 90 (July, 1996): 410–28.

Colby, Elbridge, and James F. Glass. "Legal Status of the National Guard." *Virginia Law Review* 29 (May, 1943): 839–56.

Cole, Merle T. "Martial Law in West Virginia and Major Davis as `Emperor of Tug River.'" *West Virginia History* 43 (Winter, 1982): 118–44.

———. "The Department of Special Duty Police, 1917–1918." *West Virginia History* 44 (Summer, 1983): 321–33.

———. "Organizational Development of the West Virginia State Guard." *West Virginia History* 46 (1985–86): 73–88.

———. "Maryland's State Defense Force." *Military Collector and Historian* 39 (Winter, 1987): 152–57.

———. "The Special Military Police at Morgantown Bridge, 1942–1943." *Military Affairs* 52 (Jan., 1988): 18–22.

———. "United States Guards, United States Army." *Military Collector and Historian* 40 (Spring, 1988): 2–5.

———. "Second Regiment, Infantry, Maryland State Guard: An Early State Defense Force." *Journal of Military History* 53 (Jan., 1989): 23–32.

Daugherty, Robert. "Citizen Soldiers in Peace: The Ohio National Guard, 1919–1940." Ph.D. diss., Ohio State University, 1974.

Fisher, Charles R. "The Maryland State Guard in World War II." *Military Affairs* 47 (Jan., 1983): 11–14.

Herzberg Robert. "Washington `Hot Line.'" *Combat Arms* (Mar., 1985): 12.

Jacobs, Bruce. "Memo for Record: State Military Forces after the Guard Is Called." *National Guard* 36 (June, 1982): 40.

Jacobs, Jeffrey A. "Reform of the National Guard: A Proposal to Strengthen the National Defense." *Georgetown Law Journal* 78 (Feb., 1990): 626–47.

Kendall, John M. "An Inflexible Response: United States Army Manpower Mobilization Policies, 1945–1957." Ph.D. diss., Duke University, 1982.

Kenny, Robert W. "A Brief History of the Rhode Island National Guard, 'Pride of the Ocean State.'" Army War College Library, Carlisle Barracks, Penn.

Levantrosser, William F. "The Army Reserve Merger Proposal," *Military Affairs* 30 (Fall, 1966): 135–47.

Lovett, Christopher C. "Don't You Know There's A War On? A History of the Kansas State Guard in World War II." *Kansas History* 8 (Apr., 1985–86): 226–35.

Lyons, Richard L. "The Boston Police Strike of 1919." *New England Quarterly* 20 (June, 1947): 147–68.

Miller, Margaret. "Yankee Doodle Dandies . . . or Doodle Duds?: The California Militia Awaits the Call." *California Journal* (Aug., 1992): 379–84.

"Mission of the State Guards." *Army and Navy Register* 63 (June 20, 1942): 16.

Moff, Allen A. "Citizen Militia: State Defense Forces." *American Survival Guide: The Magazine of Self Reliance.* 13 (Apr., 1991): 36–38.

Morrill, Chester, Jr. "Mission and Organization of the Army National Guard of the United States: With an Emphasis on the Period Since 1952." Ph.D. diss., American University, 1958.

Mullins, Patrick Todd. "The Militia Clause, the National Guard, and Federalism: A Constitutional Tug of War." *George Washington Law Review* 57 (Dec., 1988): 328–29.

Noble, Armond. "State Defense Forces—The New `Third Army.'" *Military* 2 (Feb., 1986): 11–12.

"Organization of State Guards." *Army and Navy Journal* 12 (Nov. 23, 1940): 327.

Pietrantoni. "True Militia: Citizen Defense Forces." *American Survival Guide*, Oct., 1985, p. 32.

Ray, Gerda. "Contested Legitimacy: Creation of the State Police in New York, 1890–1930." Ph.D. diss., University of California, Berkeley, 1990.

Sorley, Lewis. "Creighton Abrams and Active-Reserve Integration in Wartime." *Parameters* 21 (Summer, 1991): 35–50.

"State Guard Act." *Army and Navy Register* 61 (Oct. 26, 1940): 13.

"State Guard Organizations." *Army and Navy Register* 61 (Dec. 14, 1940): ?.

"State Guards." *Army and Navy Register* 62 (May 10, 1941): 8–9.

"Status of the State Guards." *Army and Navy Register* 62 (Aug. 9, 1941): 14.

"Status of the State Guards." *Army and Navy Register* 62 (Nov. 1, 1941): 5.

"State Guards." *Army and Navy Journal* (May 3, 1941): 976.

"State Guards Created." *Reserve Officer* 17 (Dec., 1940): 9.

Steeves, Kerry Ragner. "Pacific Coast Militia Rangers, 1942– 1945." M.A. thesis, University of British Columbia, 1990.

Stein, George J. "State Defense Forces: The Missing Link in National Security," *Military Review* 64 (Sept., 1984): 2–16.

Todd, Frederick P. "Our National Guard." *Military Affairs* 5 (Summer, 1941): 73–86, 152–70.

"21st Regiment, New York Guard, 1940–1945." *Military Collector and Historian* (Winter, 1983): 178–79.

Warfield, Charles Lamoreaux. "History of the Hawaii National Guard from Feudal times to June 30, 1935." M.A. thesis, University of Hawaii, 1935.

Weiner, Frederick B. "The Militia Clause of the Constitution." *Harvard Law Review* 54 (Dec., 1940): 181–220.

———. "Home Defense Organizations." *Infantry Journal* 49 (Jan., 1941): 39–40.

Westover, John Glendower. "The Evolution of the Missouri Militia, 1804–1919." Ph.D. diss., University of Missouri, 1948.

Interviews

Jerry Fogel, former president and CEO of Martin Fromm and Associates, Inc. Telephone interview, Oct. 4, 2000.

Col. Paul T. McHenry, executive director SGAUS. Interview, Charleston S.C., Oct. 21, 1995, and telephone interviews, Jan. 21, 1999, and Nov. 27, 2000.

Conference Transcripts

Annual Conference Report National Guard Association of the United States Organized 1878 [Including Addenda Executive Council Transactions 1940–1941–1942–1943] April 1,2 and 3 1943 State Capitol Harrisburg Pa. N.p., [1943].

Annual Conference Report National Guard Association of the United States Organized 1878 Sixty-Sixth Annual Convention at The Hotel Southern May 3–4–5–6 1944 Baltimore Maryland. N.p., [1944].

"Annual Meeting The Adjutants General Association April 20–21, 1942 Washington, D.C." (Photostatic copy of typed transcript.)

"Annual Meeting The Adjutants General Association April 1–3, 1943 Harrisburg, PA." (Photostatic copy of typed transcript.)

"Annual Meeting The Adjutants General Association May 3–4–5–6, 1944 Baltimore, Maryland." (Photostatic copy of typed transcript.)

"Annual Meeting The Adjutants General Association Hotel Fort Des Moines 1945 Des Moines, Iowa." (Photostatic copy of typed transcript.)

"Annual Meeting The Adjutants General Association Hotel Statler 1946 Washington, D.C." (Photostatic copy of typed transcript.)

Official Proceedings at the National Guard Association of the United States, Seventy-Third General Conference. Washington, D.C.: National Guard Association of the United States, 1951.

Official Proceedings at the National Guard Association of the United States, Seventy-Fourth General Conference. Washington, D.C.: National Guard Association of the United States, 1952.

Official Proceedings at the National Guard Association of the United States, Seventy-Fifth General Conference. Washington, D.C.: National Guard Association of the United States, 1953.

Official Proceedings at the National Guard Association of the United States, Seventy-Sixth General Conference. Washington, D.C.: National Guard Association of the United States, 1954.

Official Proceedings at the National Guard Association of the United States, Seventy-Seventh General Conference. Washington, D.C.: National Guard Association of the United States, 1955.

Proceedings of the Convention of the National Guard Association Held at Richmond, Virginia November 14 and 15 1918. N.p., [1919].

Proceedings of the Convention of the National Guard Association Of The United States Saint Louis Missouri May fifth and sixth Nineteen-nineteen. N.p., [1919].

"Special Meeting The Adjutants General Association June, 1943 Columbus, Ohio." (Photostatic copy of typed transcript.)

Index

BARRY M. STENTIFORD, an assistant professor at Grambling State University, holds a Ph.D. in history from the University of Alabama. He spent nine years as an armor officer in the Army National Guard before moving into the U.S. Army Reserve, where he holds the rank of captain.